Lóegaire Humphrey (Ed.)

Mining industry of Equatorial Guinea

Lóegaire Humphrey (Ed.)

Mining industry of Equatorial Guinea

Equatorial Guinea, Economy of Equatorial Guinea, Gross national income, Hydrocarbon exploration

Claud Press

Imprint

Publisher:
Claud Press is a trademark of
International Book Market Service Ltd., 17 Rue Meldrum, Beau Bassin, 1713-01 Mauritius
Email: info@bookmarketservice.com
Website: www.bookmarketservice.com

Published in 2011

Printed in: U.S.A., U.K., Germany. This book was not produced in Mauritius.

ISBN: 978-613-7-42097-3

Contents

Articles

References

Mining industry of Equatorial Guinea

The regulation of **mining** in **Equatorial Guinea** is handled by the Ministry of Mines, Industry, and Energy, which oversees activities in the mining and petroleum industries.[1]

Petroleum exploration and production is the core of Equatorial Guinea's economy, and accounts for over 90 percent of Equatorial Guinea's national income. As a result, little attention has been given to exploring and making use of the country's mineral resources; There is no commercial mining, although there are minor gold mining operations.[2] The Equatoguinean government has conducted preliminary explorations that they believe show potential mining opportunities for diamonds and coltan.[1] , and the Ministry of Mines is now promoting the country's mining potential in hopes of attracting investors[2]

Historically, the peoples of Equatorial Guinea produced gold and iron before being colonized by Spain. During colonial rule, there were no commercial mining operations. Following independence in 1968, Soviet and French geologists found potential deposits of gold, bauxite, tin, tungsten, and coltan. Additional surveys conducted by GEMSA, a Spanish-Equatoguinean joint venture, and UMCEG (Ocean Energy, previously named United Meridian Corporation) performed airborne and land surveys, as well as the creation of GIS databases.[3] [2]

The government claims ownership of all mineral resources, and regulates them under law 9/2006 (replacing the former Mining Law, 9/1981).[4]

Gold

Artisans using placer mining techniques retrieve gold from rivers in Rio Muni, primarily the areas of Cogo, Aconibe and Mongomo. The Equatoguinean government believes that at least 2.3 tons of gold were collectively produced from these small operations.[1]

In 2006, the United States Geological Survey reported that about 200 kg of gold were produced in Equatorial Guinea.[4]

Diamond

Indicators of diamond deposits in Gabon extend to the southern border of Rio Muni, and the Equatoguinean government believes surveys may show the potential for diamond deposits in southeastern Rio Muni.[1] [4]

Coltan

The government reports that evidence of the presence of coltan (columbite-tantalite, a tantalum ore) has been found near Aconibe and Akamiken. Very little exploration work has been performed in these areas, however.[1]

External links

- Ministry of Mines, Industry, and Energy [5] (English)

References

[1] "Hydrocarbons and Mining in Equatorial Guinea" (http://www.equatorialoil.com/pdfs/EQG Brochure August 2005.pdf) (PDF). Ministry of Mines, Industry, and Energy. . Retrieved 2008-05-01.

[2] "Equatorial Guinea: Mining" (http://www.mbendi.co.za/indy/ming/af/eq/p0005.htm). MBendi. . Retrieved 2008-05-01.

[3] null

[4] "2006 Minerals Yearbook: Equatorial Guinea" (http://minerals.usgs.gov/minerals/pubs/country/2006/myb3-2006-ek.pdf) (PDF). United States Geological Survey. . Retrieved 2008-05-01.

[5] http://www.equatorialoil.com/

Equatorial Guinea

Republic of Equatorial Guinea
República de Guinea Ecuatorial (Spanish)
République de Guinée équatoriale (French)

Motto: Unidad, Paz, Justicia (Spanish)
Unité, Paix, Justice (French)
Unity, Peace, Justice

Anthem: Caminemos pisando las sendas de nuestra inmensa felicidad
Let us walk the path of our Immense Happiness

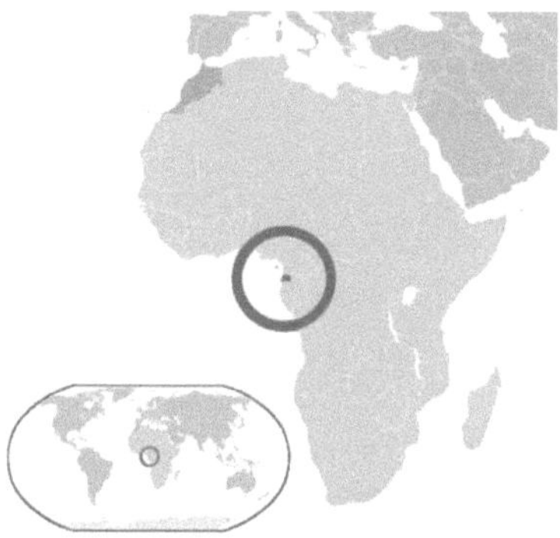

Capital		Malabo 3°45′N 8°47′E
Largest city		Bata
Official language(s)		Spanish and French
Recognised regional languages		Fang, Bube, Annobonese
Ethnic groups		85.7% Fang, 6.5% Bubi, 3.6% Mdowe, 1.6% Annobon, 1.1% Bujeba, 1.4% other (Spanish)[1]
Demonym		Equatoguinean, Equatorial Guinean
Government		Unitary presidential republic
-	President	Teodoro Obiang
-	Prime Minister	Ignacio Milam
Independence		
-	from Spain	October 12, 1968
Area		

-	Total	28,050 km^2 (144th) 10,830 sq mi
-	Water (%)	negligible
Population		
-	2009 estimate	676,000[2] (166th)
-	Density	24.1/km^2 (187th) 62.4/sq mi
GDP (PPP)		2010 estimate
-	Total	$24.146 billion[3]
-	Per capita	$34,824[3] (22nd)
GDP (nominal)		2010 estimate
-	Total	$14.006 billion[3]
-	Per capita	$20,200[3]
HDI (2010)		▲ 0.538[4] (medium) (117th)
Currency		Central African CFA franc (XAF)
Time zone		WAT (UTC+1)
-	Summer (DST)	not observed (UTC+1)
Drives on the		right
ISO 3166 code		GQ
Internet TLD		.gq
Calling code		240

Equatorial Guinea, officially the **Republic of Equatorial Guinea** (Spanish: *República de Guinea Ecuatorial*, pronounced [reˈpuβlika ðe ɣiˈne.a ekwatoˈɾjal]; French: *République de Guinée équatoriale*, pronounced: [ʁepyblik də ɡine ekwatoʁjal]) is a country located in Middle Africa. It comprises two parts: a Continental Region (Río Muni), including several small offshore islands like Corisco, Elobey Grande and Elobey Chico; and an insular region containing Annobón island and Bioko island (formerly *Fernando Po*) where the capital Malabo is situated.

Annobón is the southernmost island of Equatorial Guinea and is situated just south of the equator. Bioko island is the northernmost point of Equatorial Guinea. Between the two islands and to the east is the mainland region. Equatorial Guinea is bordered by Cameroon on the north, Gabon on the south and east, and the Gulf of Guinea on the west, where the island nation of São Tomé and Príncipe is located between Bioko and Annobón. Formerly the colony of Spanish Guinea, its post-independence name is suggestive of its location near both the equator and the Gulf of Guinea. Besides the Spanish cities of Ceuta and Melilla on the Mediterranean coast next to Morocco, it is the only territory in mainland Africa with Spanish as the official language.

With an area of 28000 square kilometres (11000 sq mi) Equatorial Guinea is one of the smallest countries in continental Africa. It is also the richest per capita [5] ; however, the wealth is distributed very unevenly, with 70% of the population living under the United Nations Poverty Threshold of $2/day. With a population of 650,702, Equatorial Guinea is the third smallest country in continental Africa.[6] It is also the second smallest United Nations (UN) member from continental Africa.

The discovery of sizeable petroleum reserves in recent years is altering the economic and political status of the country. Its gross domestic product (GDP) per capita ranks 28th in the world;[7] however, most of the country's considerable oil wealth actually lies in the hands of only a few people.

Equatorial Guinea has one of the worst human rights records in the world, consistently ranking among the "worst of the worst" in Freedom House's annual survey of political and civil rights[8] and Reporters Without Borders ranks President Obiang among its "predators" of press freedom.[9] Out of 44 sub-Saharan countries, Equatorial Guinea ranks 9th highest in the Human Development Index (HDI) and 115th overall, which is among the *medium HDI* countries.

Geography

Equatorial Guinea is located in west central Africa. The country consists of a mainland territory, Río Muni, which is bordered by Cameroon to the north and Gabon to the east and south and five small islands, Bioko, Corisco, Annobón, Small Elobey and Great Elobey. Bioko, the site of the capital, Malabo, lies about 40 kilometers (25 mi) off the coast of Cameroon. Annobón island is about 350 kilometers (220 mi) west-south-west of Cape Lopez in Gabon. Corisco and the two Elobey islands are in Corisco Bay, on the border of Río Muni and Gabon.

Equatorial Guinea lies between latitudes 4°N and 2°S, and longitudes 5° and 12°E. Despite its name, no part of the country's territory lies on the equator - it is entirely in the northern hemisphere, except for the island province of Annobón which is about 155 km south of the equator.

Equatorial Guinea spans several ecoregions. Río Muni region lies within the Atlantic Equatorial coastal forests ecoregion except for patches of Central African mangroves on the coast, especially in the Muni River estuary. The Cross-Sanaga-Bioko coastal forests ecoregion covers most of Bioko and as well as the adjacent portions of Cameroon and Nigeria on the African mainland, and the Mount Cameroon and Bioko montane forests ecoregion covers the highlands of Bioko and nearby Mount Cameroon.

The São Tomé, Príncipe, and Annobón moist lowland forests ecoregion covers all of Annobón, as well as São Tomé and Príncipe.

Climate

Equatorial Guinea has a tropical climate with distinct wet and dry seasons. From June to August, Río Muni is dry and Bioko wet; from December to February, the reverse occurs. In between there is gradual transition. Rain or mist occurs daily on Annobón, where a cloudless day has never been registered. The temperature at Malabo, Bioko, ranges from 16 °C (61 °F) to 33 °C (91 °F), though on the southern Moka Plateau normal high temperatures are only 21 °C (70 °F). In Río Muni, the average temperature is about 27 °C (81 °F). Annual rainfall varies from 1930 mm (76 in) at Malabo to 10920 mm (430 in)) at Ureka, Bioko, but Río Muni is somewhat drier.[10]

History

In the continental region that is now Equatorial Guinea there are believed to have been pygmies, of whom only isolated pockets remain in northern Río Muni. Bantu migrations between the 17th and 19th centuries brought the coastal tribes and later the Fang. Elements of the latter may have generated the Bubi, who emigrated to Bioko from Cameroon and Rio Muni in several waves and succeeded former Neolithic populations. The Annobón population, native to Angola, was introduced by the Portuguese via São Tomé island (São Tomé and Príncipe).

The Portuguese explorer Fernão do Pó, seeking a path to India, is credited as being the first European to discover the island of Bioko in 1472. He called it *Formosa* ("Beautiful"), but it quickly took on the name of its European discoverer. The islands of Fernando Pó and Annobón were colonized by Portugal in 1474.

In 1778, the island, adjacent islets, and commercial rights to the mainland between the Niger and Ogoue Rivers were ceded to Spain in exchange for territory in the American continent (Treaty of El Pardo, between Queen Maria I of Portugal and King Charles III of Spain). Between 1778 and 1810, the territory of Equatorial Guinea depended administratively on the Viceroyalty of the Río de la Plata, with seat in Buenos Aires.

From 1827 to 1843, the United Kingdom established a base on the island to combat the slave trade,[11] which was then moved to Sierra Leone upon agreement with Spain in 1843. In 1844, on restoration of Spanish sovereignty, it became known as the *Territorios Españoles del Golfo de Guinea Ecuatorial*. The mainland portion, Rio Muni, became a protectorate in 1885 and a colony in 1900. Conflicting claims to the mainland were settled by the Treaty of Paris in 1900, and periodically, the mainland territories were united administratively under Spanish rule. Between 1926 and 1959 they were united as the colony of Spanish Guinea.

In September 1968, Francisco Macías Nguema was elected first president of Equatorial Guinea, and independence was recognised on October 12, 1968. In July 1970, Nguema created a single-party state. Nguema's reign of terror led to the death or exile of up to 1/3 of the country's population. Out of a population of 300,000, an estimated 80,000 had been killed.[12] [13] The economy collapsed, and skilled citizens and foreigners left.[14] Teodoro Obiang deposed Francisco Macías Nguema on August 3, 1979, in a bloody coup d'état.

Politics

The current president of Equatorial Guinea is Teodoro Obiang Nguema Mbasogo. The 1982 constitution of Equatorial Guinea, written following the 1979 deposition of dictator Francisco Macías Nguema and with help from the UN, gives the presidency extensive powers, including naming and dismissing members of the cabinet, making laws by decree, dissolving the Chamber of Representatives, negotiating and ratifying treaties and serving as commander in chief of the armed forces. The Prime Minister, Ignacio Milam Tang is appointed by the President and operates under powers designated by the President.

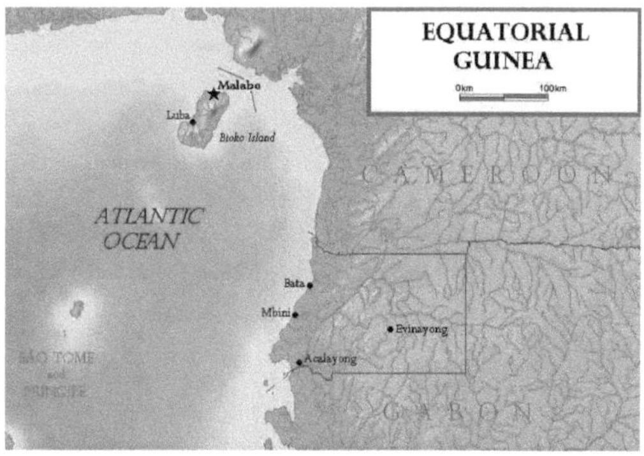

Map of Equatorial Guinea

On Christmas 1975, Macías had 150 alleged coup plotters executed to the sound of a band playing Mary Hopkin's tune *Those Were the Days* in a national stadium.[15] It is estimated that 100,000 people (approximately one-third of the population) were killed or fled into exile during Macías' reign.[16]

President Obiang overthrew previous dictator Francisco Macías Nguema on 3 August 1979 in a bloody coup d'état. Since August 1979 some 12 real and perceived unsuccessful coup attempts have occurred. The 'real' coup attempts were often perpetrated in an attempt by rival elites to seize the state's economic resources.[16]

Under President Obiang, the U.S. Agency for International Development (USAID) entered into a Memorandum of Understanding (MOU) with the Republic of Equatorial Guinea, in April 2006, to establish a Social Development Fund in the country, implementing projects in the areas of health, education, women's affairs and the environment.[17]

Since 2005, MPRI, a U.S. based international policing consulting company, has worked in Equatorial Guinea to train police forces in appropriate human rights practices. In February 2010, Equatorial Guinea signed a contract with the MPRI subsidiary of the US defense corporation, L3 Communications for coastal surveillance and maritime security in the Gulf of Guinea.[18] [19]

Although President Obiang signed a national anti-torture decree in 2006 to ban all forms of abuse and improper treatment in Equatorial Guinea and commissioned the renovation and modernization of Black Beach prison in 2007 to ensure the humane treatment of prisoners,[20] human rights abuses continue. Human Rights Watch and Amnesty

International among other non-governmental organizations have documented severe human rights abuses in prisons, including torture, beatings, unexplained deaths and illegal detention.[21] [22]

Under President Obiang, the basic infrastructure of Equatorial Guinea has also improved. Asphalt now covers more than 80% of the national roads and ports and airports are being built across the entire country.[23]

According to a March 2004 BBC profile,[24] politics within the country are currently dominated by tensions between Obiang's son, Teodorin, and other close relatives with powerful positions in the security forces. The tension may be rooted in power shift arising from the dramatic increase in oil production which has occurred since 1997.

A November 2004 report[25] named Mark Thatcher as a financial backer of a 2004 Equatorial Guinea coup d'état attempt to topple Obiang, organized by Simon Mann. Various accounts also name the United Kingdom's MI6, the United States' CIA, and Spain as having been tacit supporters of the coup attempt.[26] Nevertheless, the Amnesty International report released in June 2005[27] on the ensuing trial of those allegedly involved highlighted the prosecution's failure to produce conclusive evidence that a coup attempt had actually taken place.

Simon Mann was released from prison on November 3, 2009 for humanitarian reasons. The presidential decree pardoning Mann from prison cites concerns about his physical health and the need for him to receive ongoing care in his home country.[28]

President Obiang was re-elected to serve an additional term in 2009 in an election deemed by the African Union as "in line with electoral law".[29] The President reappointed Prime Minister Ignacio Milam Tang and installed a new government in Equatorial Guinea on January 12, 2010.[30]

The new government is dedicated to strengthening the "cooperation and friendship" with the Barack Obama administration. During a meeting on the sidelines of the recent United Nations General Assembly, President Obiang urged President Obama to institute a U.S–Africa summit, to strengthen the cooperation between the United States and Africa.[23]

Administrative divisions

Equatorial Guinea is divided into seven provinces (capitals appear in parentheses):

1. Annobón Province (San Antonio de Palé)
2. Bioko Norte Province (Malabo)
3. Bioko Sur Province (Luba)
4. Centro Sur Province (Evinayong)
5. Kié-Ntem Province (Ebebiyín)
6. Litoral Province (Bata)
7. Wele-Nzas Province (Mongomo)

The provinces are further divided into districts.

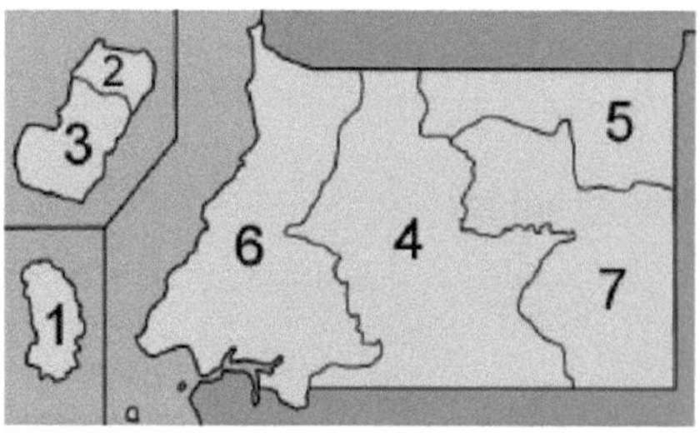

Provinces of Equatorial Guinea

Economy

Pre-independence Equatorial Guinea counted on cocoa production for hard currency earnings. It had the highest per capita income of Africa in 1959. On January 1, 1985, the country became the first non-Francophone African member of the franc zone, adopting the CFA as its currency. The national currency, the ekwele, was previously linked to the Spanish peseta.[31]

The discovery of large oil reserves in 1996 and its subsequent exploitation have contributed to a dramatic increase in government revenue. As of 2004,[32] Equatorial Guinea is the third-largest oil producer in Sub-Saharan Africa. Its oil production has risen to 360000 barrels per day (57000 m^3/d), up from 220,000 only two years earlier.

Forestry, farming, and fishing are also major components of GDP. Subsistence farming predominates. The deterioration of the rural economy under successive brutal regimes has diminished any potential for agriculture-led growth.

In July 2004, the United States Senate published an investigation into Riggs Bank, a Washington-based bank into which most of Equatorial Guinea's oil revenues were paid until recently, and which also banked for Chile's Augusto Pinochet. The Senate report, as to Equatorial Guinea, showed that at least $35 million were siphoned off by Obiang, his family and senior officials of his regime. The president has denied any wrongdoing. While Riggs Bank in February 2005 paid $9 million as restitution for its banking for Chile's Augusto Pinochet, no restitution was made with regard to Equatorial Guinea, as reported in detail in an Anti-Money Laundering Report from Inner City Press.[33]

Equatorial Guinea is a member of the Organization for the Harmonization of Business Law in Africa (OHADA).[34]

Equatorial Guinea tried to become validated as an Extractive Industries Transparency Initiative (EITI) Compliant country, working toward transparency in reporting of oil revenues and the prudent use of natural resource wealth. The country was one of 30 Candidate countries and obtained Candidate status February 22, 2008. It was then were required to meet a number of obligations to do so, including committing to working with civil society and companies on EITI implementation, appointing a senior individual to lead on EITI implementation, and publishing a fully costed Work Plan with measurable targets, a timetable for implementation and an assessment of capacity constraints. However, when Equatorial Guinea applied to extend the deadline for completing EITI validation, the EITI Board did not agree to grant Equatorial Guinea an extension.[35]

Demographics

Equatorial Guinean children of Bubi descent.

The majority of the people of Equatorial Guinea are of Bantu origin. The largest tribe, the Fang, is indigenous to the mainland, but substantial migration to Bioko Island has resulted in the Fang population exceeding that of the earlier Bantu inhabitants. The Fang constitute 80% of the population[36] and comprise 67 clans. Those in the northern part of Rio Muni speak Fang-Ntumu, while those in the south speak Fang-Okah; the two dialects have differences but are mutually intelligible. Dialects of Fang are also spoken in parts of neighboring Cameroon (Bulu) and Gabon. These dialects, while still intelligible, are more distinct. The Bulu Fang of Cameroon were traditional rivals of Fang in Rio Muni. The Bubi, who constitute 15% of the population, are indigenous to Bioko Island. The traditional demarcation line between Fang and beach tribes was the village of Niefang (limit of the Fang) inland from Bata.

In addition, there are coastal tribes, sometimes referred to as Ndowe or "Playeros" (*Beach People* in Spanish): Combes, Bujebas, Balengues, and Bengas on the mainland and small islands, and Fernandinos, a Krio community on Bioko Island. Together, these groups compose 5% of the population. Some Europeans (largely of Spanish or

Portuguese descent) – among them mixed with African ethnicity – also live in the nation. Most Spaniards left after independence. There is a growing number of foreigners from neighboring Cameroon, Nigeria, and Gabon. Equatorial Guinea received Asians and black Africans from other countries as workers on cocoa and coffee plantations. Other black Africans came from Liberia, Angola, and Mozambique. Most of the Asian population is Chinese, with small numbers of Indians.

Equatorial Guinea also allowed many fortune-seeking European settlers of other nationalities, including British, French and Germans. There is also a group of Israelis, which are employed at the Centro Médico La Paz in Bata. After independence, thousands of Equatorial Guineans went to Spain. Another 100,000 Equatorial Guineans went to Cameroon, Gabon, and Nigeria because of the dictatorship of Francisco Macías Nguema. Some Equatorial Guinean communities are also to be found in Latin America, the United States, Portugal, and France. Oil extraction has contributed to a doubling of the population in Malabo.

Religion in Equatorial Guinea	
religion	percent
Christian	93%
Muslim	1%
Other (indigenous beliefs/Baha'i)	6%

Religion

The principal religion in Equatorial Guinea is Christianity which is the faith of 93% of the population. These are predominately Roman Catholic (87%) while a minority are Protestants (5%). Another 5% of the population follow indigenous beliefs and the final 2% comprises Muslims, Bahá'í Faith, and other beliefs.[37]

Languages

The Constitutional Law that amends article four of the Fundamental Law of the State establishes that the official languages of the Republic of Equatorial Guinea are Spanish and French. In July 2007, President Teodoro Obiang Nguema announced his government's decision for Portuguese to become Equatorial Guinea's third official language, in order to meet one of the requirements to apply for full membership in the Community of Portuguese Language Countries (CPLP), the other one being political reforms allowing for effective democracy and the respect for human rights. This upgrading from its current Associate Observer condition would result in Equatorial Guinea being able to access several professional and academic exchange programs and the facilitation of cross-border circulation of citizens. Its application is currently being assessed by other CPLP members.[38]

According to a few sources Portuguese is an official language now,[38] [39] [40] [41] [42] but many other sources, including official Equatorial Guinean sources, do not treat Portuguese as an official language, which indicates that the decision above was not implemented.[43] [44] [45] [46] [47] [48] [49] During the VIII Conference of the Community of Portuguese Language Countries in Luanda (23 July 2010), Equatorial Guinea formally requested the membership in this organization. The final declaration of the conference stated that one of the conditions is the adoption of Portuguese as an official language of Equatorial Guinea, which clearly indicates that by the time of the conference the Portuguese language was not an official language of Equatorial Guinea.[50] However after finalizing the last CPLP summit, the President of Equatorial Guinea, in declarations to the international media, stated that his country maintains the hope of forming part of this community, with which Equatorial Guinea, for its history and its culture, is very close. He told that on the 20 July 2010, the Government of Equatorial Guinea had approved the decree by which the Portuguese language became the third official language of the country, awaiting only its approval in the House of People's Representatives.[51] [52] So far no official confirmation of approving the decree by the Parliament – official languages are listed in the constitution, so Parliament should adopt the constitutional law that amends article four of the constitulion. Aboriginal languages are recognized as integral parts of the national culture" (Constitutional Law No. 1/1998 of January 21). The great majority of Equatorial Guineans speak Spanish,[38] especially those living in the capital, Malabo. Spanish has been an official language since 1844.

Languages spoken in the country are Spanish, Equatoguinean Spanish, native languages (including Fang, Bube, Benga, Pichinglis, Ndowe, Balengue, Bujeba, Bissio, Gumu, Igbo and a nearly extinct Baseke), French, others (mainly English, or German, as well as Annobonese language (Fá d'Ambô) a Portuguese creole, and Fernando Poo Creole English).

Culture

In June 1984, the First Hispanic-African Cultural Congress was convened to explore the cultural identity of Equatorial Guinea. The congress constituted the center of integration and the marriage of the Hispanic culture with African cultures.[31]

Education

Under the regime of Francisco Macias, education had been significantly neglected with few children receiving any type of education. Under President Obiang, the illiteracy rate dropped from 73 percent to 13 percent [53] and the number of primary school students has risen from 65,000 in 1986 to more than 100,000 in 1994. Education is free and compulsory for children between the ages of 6 and 14.[31]

The Equatorial Guinea government has also partnered with Hess Corporation and The Academy for Educational Development (AED) to establish a $20 million education program through which primary school teachers participate in a training program to teach modern child development techniques.[54]

In recent years, with change in economic/political climate and government social agendas, several cultural dispersion and literacy organizations are now located in the country, founded chiefly with the financial support of the Spanish government. The country has one university, the Universidad Nacional de Guinea Ecuatorial (UNGE) with a campus in Malabo and a Faculty of Medicine located in Bata on the mainland. In 2009 the university produced the first 110 national doctors.[23] The Bata Medical School is supported principally by the government of Cuba and staffed by Cuban medical educators and physicians, however, it is predicted that Equatorial Guinea will have enough national doctors in the country to be self-sufficient within the next five years.[23]

Health

Equatorial Guinea's innovative malaria control programs have had a remarkable impact on malaria infection, disease, and mortality in the population.[55] Their program consists of twice-yearly indoor residual spraying (IRS), the introduction of artemisinin combination treatment (ACTs), the use of intermittent preventive treatment in pregnant women (IPTp) and the introduction of very high coverage with long-lasting insecticide treated mosquito nets (LLINs). The result of their efforts resulted in a reduction in all-cause under-five mortality from 152 to 55 deaths per 1,000 live births (down 64%); and the drop occurred rapidly and timed directly with the beginning of the program.[56]

Malaria

In 2011, the advocacy group Roll Back Malaria reported that companies working in Equatorial Guinea "have worked to prevent malaria among their workers and workers' dependents and have seen an excellent return on investment, with significant reductions in malaria-related illnesses and deaths, worker absenteeism, and malaria-related spending."[57]

Transportation

Air transport

Every airline registered in the country appears on the list of air carriers prohibited in the European Union (EU) which means that it is banned for safety reasons from operating services of any kind within the EU.[58]

Due to the big-oil presence in the country, internationally recognised carriers fly to Malabo (Bioko). The carriers include:

- Iberia - from Madrid
- Air Europa - from Madrid
- Air France - from Paris
- Lufthansa - from Frankfurt
- Kenya Airways - from Nairobi
- Danish Air Transport - Internal and charters

Communications

The principal means of communication within the country are three state-operated FM radio stations; there are also five shortwave radio stations. There are two newspapers and two magazines. Television Nacional, the television network, is state operated.[59] [60]

Most of the media companies practice heavy self-censorship, and are banned by law from criticising public figures. The state-owned media and the main private radio station are under the directorship of Teodorin Nguema Obiang, the president's son.

Landline telephone penetration is low, with only two lines available for every 100 persons.[60] There is one GSM mobile telephone operator, with coverage of Malabo, Bata, and several mainland cities.[61] [62] As of 2009, approximately forty percent of the population subscribed to mobile telephone services.[63] The only telephone provider in Equatorial Guinea is Orange.

Equatorial Guinea has nine (as of 2009) Internet service providers, which serves more than 8,000 users.[60]

Sports

Football

Further information: Equatorial Guinea national football team and Equatorial Guinea women's national football team

Equatorial Guinea has been chosen to co-host the 2012 African Cup of Nations in partnership with Gabon. The country was also chosen to host the 2008 Women's African Football Championship, which they won. The Women's National Team qualified for the 2011 World Cup in Germany.

Swimming

Equatorial Guinea is also famous for the national swimming champion Eric Moussambani, nicknamed "Eric the Eel".

In fiction

Frederick Forsyth's 1974 novel *The Dogs of War* is set in the fictional platinum-rich 'Republic of Zangaro', which is based on Equatorial Guinea. There is also a 1981 film adaptation of the book, also called The Dogs of War.

Fernando Po (now Bioko) is featured prominently in the 1975 science fiction work *The Illuminatus! Trilogy* by Robert Shea and Robert Anton Wilson. The island (and, in turn, the country) experience a series of coups in the story which lead the world to the verge of nuclear war. The story also hypothesizes that Fernando Po is the last remaining piece of the sunken continent of Atlantis.

Most of the action in the American novelist Robin Cook's book, *Chromosome 6*, takes place at a primate research facility based in Equatorial Guinea due to the country's permissive laws. The book also discusses some of the geography, history, and peoples of the country.

Episode 2 of the British sitcom Yes Minister, The Official Visit, situates the fictional *lesser developed country* of Buranda in what is actually Equatorial Guinea.

In the novel Limit (2009) by Frank Schätzing, which takes place in 2025, the country's history (and future history) plays a significant role in the plot.

See also

- Bight of Bonny also known as the Bight of Biafra
- Cameroon line
- Gulf of Guinea
- Telecommunications in Equatorial Guinea
- Foreign relations of Equatorial Guinea
- Military of Equatorial Guinea
- Scouting in Equatorial Guinea
- Transport in Equatorial Guinea
- 2004 Equatorial Guinea coup d'état attempt
- Equatoguinean literature in Spanish
- List of cities in Equatorial Guinea

Notes and references

This article incorporates public domain material from websites or documents [64] of the CIA World Factbook.

[1] Cia World Factbook; Equatorial Guinea (https://www.cia.gov/library/publications/the-world-factbook/geos/ek.html)

[2] Department of Economic and Social Affairs Population Division (2009) (PDF). *World Population Prospects, Table A.1* (http://www.un.org/esa/population/publications/wpp2008/wpp2008_text_tables.pdf). 2008 revision. United Nations. . Retrieved 2009-03-12.

[3] World Development Indicators database (http://data.worldbank.org/country/equatorial-guinea), World Bank, accessed on August 23, 2011.

[4] "Human Development Report 2009. Human development index trends: Table G" (http://hdr.undp.org/en/media/HDR_2009_EN_Complete.pdf) (PDF). The United Nations. . Retrieved 2009-10-10.

[5] http://www.indexmundi.com/g/r.aspx?c=mr&v=67

[6] Seychelles, The Gambia, Djibouti, Rwanda, Burundi, Cape Verde, Comoros, Swaziland, and São Tomé and Príncipe are smaller in terms of area, and Djibouti and the Sahrawi Arab Democratic Republic have smaller populations, although the population of the latter is disputed

[7] **CIA World Factbook**: GDP - per capita (PPP) (https://www.cia.gov/library/publications/the-world-factbook/rankorder/2004rank.html), The World Factbook, Central Intelligence Agency.

[8] http://freedomhouse.org/uploads/special_report/88.pdf

[9] http://en.rsf.org/equatorial-guinea.html

[10] Nations Encyclopedia (http://www.nationsencyclopedia.com/Africa/Equatorial-Guinea-CLIMATE.html)

[11] See Encyclopedia Britannica, 1911, "Fernando Po"

[12] Sengupta, Kim (2007-05-11). "Coup plotter faces life in Africa's most notorious jail" (http://news.independent.co.uk/world/africa/article2530772.ece). London: News.independent.co.uk. . Retrieved 2010-05-03.

[13] "True hell on earth: Simon Mann faces imprisonment in the cruellest jail on the planet" (http://www.dailymail.co.uk/pages/live/articles/news/news.html?in_article_id=455635&in_page_id=1770). London: Dailymail.co.uk. 2007-05-18. . Retrieved 2010-05-03.

[14] Daniels, Anthony (2004-08-29). "If you think this one's bad you should have seen his uncle" (http://www.telegraph.co.uk/opinion/main.jhtml?xml=/opinion/2004/08/29/do2902.xml&sSheet=/opinion/2004/08/29/ixopinion.html). London: Telegraph.co.uk. . Retrieved 2010-05-03.

[15] "Oil Gives African Nation a Chance for Change" (http://www.hartford-hwp.com/archives/35/238.html). The Washington Post. 2001-05-13. .

[16] Alex Vines (9 July 2009). "Well Oiled" (http://www.hrw.org/en/node/84252/section/3). Human Rights Watch. . Retrieved 21 January 2011.

[17] Heather Layman, LPA. "USAID" (http://www.usaid.gov/press/releases/2006/pr060411_2.html). Usaid.gov. . Retrieved 2010-05-03.

[18] http://jutiagroup.com/2010/04/05/tension-builds-in-the-gulf-of-guinea-as-competition-for-economic-resources-increases/

[19] http://blog.taragana.com/business/2010/02/24/13-communications-coast-surveillance-contract-with-equatorial-guinea-could-be-worth-250m-34925/

[20] (http://www.mpri.com/esite/index.php/content/services/organizational_reform_amp_institutional_capacity_building/) MPRI

[21] http://www.amnesty.org/en/region/equatorial-guinea/report-2009

[22] http://www.hrw.org/en/world-report-2009/equatorial-guinea

[23] (http://www1.voanews.com/english/news/africa/Equatorial-Guinea-Minister-Seeks-Strong-Ties-With-US-89956902.html) VOA News

[24] Shaxson, Nicholas (2004-03-17). "Profile: Equatorial Guinea's great survivor" (http://news.bbc.co.uk/1/hi/world/africa/3516588.stm). BBC News. .

[25] "Thatcher faces 15 years in prison" (http://www.smh.com.au/articles/2004/08/26/1093518010003.html). The Sydney Morning Herald. 2004-08-27. .

[26] MacKay, Neil (2004-08-29). "The US knew, Spain knew, Britain knew. Whose coup was it?" (http://findarticles.com/p/articles/mi_qn4156/is_20040829/ai_n12590798/?tag=content;col1). Sunday Herald. .

[27] "Equatorial Guinea, A trial with too many flaws" (http://web.amnesty.org/library/Index/ENGAFR240052005?open&of=ENG-GNQ). Amnesty International. 2005-06-07. .

[28] "Presidential Decree" (http://www.republicofequatorialguinea.net/news/?newsID=730). Republicofequatorialguinea.net. . Retrieved 2010-05-03.

[29] Factoria Audiovisual S.R.L.. "Declaración de la Unión Africana, sobre la supervisión de los comicios electorales - Página Oficial de la Oficina de Información y Prensa de Guinea Ecuatorial" (http://www.guineaecuatorialpress.com/noticia.php?id=259). Guineaecuatorialpress.com. . Retrieved 2010-05-03.

[30] "UPDATE 1-Tang renamed as Equatorial Guinea PM | News by Country | Reuters" (http://af.reuters.com/article/equatorialGuineaNews/idAFLDE60B26H20100112). Af.reuters.com. 2010-01-12. . Retrieved 2010-05-03.

[31] "EquatorialGuinea.org; Retrieved 12/08/1998" (http://web.archive.org/web/19991003081314/http://www.equatorialguinea.org/information1387/information.htm). Web.archive.org. Archived from the original (http://www.equatorialguinea.org/information1387/information.htm) on 1999-10-03. . Retrieved 2010-05-03.

[32] Justin Blum (September 7, 2004). "U.S. Oil Firms Entwined in Equatorial Guinea Deals" (http://www.washingtonpost.com/wp-dyn/articles/A1101-2004Sep6.html). washingtonpost.com. . Retrieved 2008-07-09.

[33] "Inner City Press / Finance Watch: "Follow the Money, Watchdog the Regulators"" (http://www.innercitypress.org/finwatch.html). Innercitypress.org. . Retrieved 2010-05-03.

[34] "OHADA.com: The business law portal in Africa" (http://www.ohada.com/index.php). . Retrieved 2009-03-22

[35] (http://www.eitransparency.org/EquatorialGuinea) Extractive Industries Transparency Initiative

[36] "Equatorial Guinea's God". *BBC*. 26 July 2003.

[37] "U.S. Department of State" (http://www.state.gov/g/drl/rls/irf/2007/90095.htm). State.gov. 2007-09-14. . Retrieved 2010-05-03.

[38] "Obiang convierte al portugués en tercer idioma oficial para entrar en la Comunidad lusófona de Naciones" (http://actualidad.terra.es/internacional/articulo/obiang_comunidad_naciones_1710388.htm), *Terra*. 13-07-2007 *(translate* (http://translate.google.com/translate?u=http://actualidad.terra.es/internacional/articulo/obiang_comunidad_naciones_1710388.htm&hl=en&langpair=es|en&tbb=1&ie=windows-1252)*)*

[39] Europa Press: Portuguese becomes E.G. 3rd official language (http://www.europapress.es/epsocial/noticia-portugues-convierte-tercer-idioma-oficial-guinea-ecuatorial-pocos-dias-cumbre-cplp-20100721152101.html) (In Spanish)

[40] Countries by Languages (http://www.nationsonline.org/oneworld/countries_by_languages.htm) (See Portuguese)

[41] The Portuguese Language (http://www.buzzle.com/articles/the-portuguese-language.html)

[42] PORTUGUESE LANGUAGE STATISTICS (http://www.portugueselanguageguide.com/portuguese/facts/stats)

[43] Oficina de Información y Prensa de Guinea Ecuatorial, Ministerio de Información, Cultura y Turismo (http://www.guineaecuatorialpress.com/noticia.php?id=134): *El Español es la lengua oficial administrativa y de enseñanza. El francés es la segunda lengua oficial y casi todas las etnias hablan las denominadas lenguas bantúes.*

[44] Embassy of Equatorial Guinea in London (http://www.embarege-londres.org/aboutGE.cfm)

[45] US Department of State – Background Note: Equatorial Guinea (2010) (http://www.state.gov/r/pa/ei/bgn/7221.htm#people)

[46] United Nations list of country names (May 2011) (http://unstats.un.org/unsd/geoinfo/UNGEGN/docs/26th-gegn-docs/WP/WP54_UNGEGN WG Country Names Document 2011.pdf)

[47] French government list of country names (2010) (http://www.cnig.gouv.fr/Front/docs/cms/cntpvm_enrichi_13avril2010_127540209260628000.pdf)

[48] Polish list of country names (2007) (http://ksng.gugik.gov.pl/pliki/nazwy_panstw_swiata.pdf) with 2009 update (http://ksng.gugik.gov.pl/pliki/nazwy_panstw-aktualizacja_maj_2009.pdf)

[49] CIA The World Factbook (2010) (https://www.cia.gov/library/publications/the-world-factbook/geos/ek.html)

[50] "Declaração de Luanda" of the VIII Conferência de Chefes de Estado e de Governo da Comunidade dos Países de Língua Portuguesa (http://www.cplp.org/Admin/Public/DWSDownload.aspx?File=/Files/Filer/cplp/CCEG/VIII_CCEG/PDFs/Docs+finais+cimeira/Dec_Luanda_act13ago.pdf)

[51] (http://www.guineaecuatorialpress.com/noticia.php?id=712)

[52] http://www.guineaecuatorialpress.com/noticia.php?id=703

[53] (https://www.cia.gov/library/publications/the-world-factbook/geos/ek.html) CIA Factbook

[54] (http://www.aed.org/News/Stories/equatorial_guinea_partnership.cfm) AED

[55] (http://www.ajtmh.org/cgi/content/full/80/6/879#R1) Good News in Malaria Control

[56] (http://www.ajtmh.org/cgi/content/abstract/80/6/882?ijkey=42e57e2ed496ad1cc91ad3c34a8636edd294d458&keytype2=tf_ipsecsha) Marked Increase in Child Survival after Four Years of Intensive Malaria Control

[57] (http://www.rollbackmalaria.org/ProgressImpactSeries/report6.html) Roll Back Malaria Progress & Impact Series: Business Investing in Malaria Control: Economic Returns and a Healthy Workforce for Africa

[58] List of banned E.U. air carriers (http://ec.europa.eu/transport/air-ban/pdf/list_en.pdf)

[59] "Country Profile: Equatorial Guinea: Media" (http://news.bbc.co.uk/2/hi/africa/country_profiles/1023151.stm#media). BBC News. 2008-01-26. .

[60] "CIA World Factbook: Equatorial Guinea; Communications" (https://www.cia.gov/library/publications/the-world-factbook/geos/ek.html). Central Intelligence Agency. 2008-03-20. .

[61] "GSMWorld Providers: Equatorial Guinea" (http://www.gsmworld.com/roaming/gsminfo/cou_gq.shtml). GSM World. 2008. .

[62] "GSMWorld GETESA Coverage Map" (http://www.gsmworld.com/cgi-bin/ni_map.pl?cc=gq&net=ge). GSM World. 2008. .

[63] "CIA World Factbook: Equatorial Guinea" (https://www.cia.gov/library/publications/the-world-factbook/geos/ek.html). Central Intelligence Agency. 2009-02-09. .

[64] https://www.cia.gov/library/publications/the-world-factbook/index.html

Books

- Max Liniger-Goumaz, *Small is not Always Beautiful: The Story of Equatorial Guinea* (French 1986, translated 1989) ISBN 0-389-20861-2
- Ibrahim K. Sundiata, *Equatorial Guinea: Colonialism, State Terror, and the Search for Stability* (1990, Boulder: Westview Press) ISBN 0-8133-0429-6
- Robert Klitgaard. 1990. *Tropical Gangsters*. New York: Basic Books. (World Bank economist tries to assist pre-oil Equatorial Guinea) ISBN 0-465-08760-4
- D.L. Claret. *Cien años de evangelización en Guinea Ecuatorial (1883–1983)/ One Hundred Years of Evangelism in Equatorial Guinea* (1983, Barcelona: Claretian Missionaries)
- Adam Roberts, *The Wonga Coup: Guns, Thugs and a Ruthless Determination to Create Mayhem in an Oil-Rich Corner of Africa* (2006, PublicAffairs) ISBN 1-58648-371-4

External links

Government

- Official Government of Equatorial Guinea website (http://guinea-equatorial.com/) (Spanish, English)
- Honorary Consulate of Equatorial Guinea in Romania (http://www.ecaligiuri.com/index2.php) (Spanish, Romanian, English)
- Chief of State and Cabinet Members (https://www.cia.gov/library/publications/world-leaders-1/world-leaders-e/equatorial-guinea.html)

General information

- Country Profile (http://news.bbc.co.uk/1/hi/world/africa/country_profiles/1023151.stm) from BBC News
- Equatorial Guinea (https://www.cia.gov/library/publications/the-world-factbook/geos/ek.html) entry at *The World Factbook*
- Equatorial Guinea (http://ucblibraries.colorado.edu/govpubs/for/equatorialguinea.htm) from *UCB Libraries GovPubs*
- Equatorial Guinea (http://www.dmoz.org/Regional/Africa/Equatorial_Guinea/) at the Open Directory Project
- Wikimedia Atlas of Equatorial Guinea
- Topographical map (http://www.iradier.org/colaboracion5.htm) (Spanish)
- Equatorial Guinea travel guide from Wikitravel

News media

- Equatorial Guinea news headline links (http://allafrica.com/equatorialguinea/) from AllAfrica.com
- Guinea-Ecuatorial.net (http://www.guinea-ecuatorial.net/ms/main.asp) (Spanish, some French)

Languages

- Las lenguas de Guinea Ecuatorial: Materiales bibliográficos para su estudio (http://cvc.cervantes.es/lengua/thesaurus/pdf/39/TH_39_123_190_0.pdf) (in Spanish)

Other

- Spanish Ministry's Plan for Africa 2006-2008 (http://www.maec.es/es/Home/Paginas/planafrica2006_08.aspx) - (Spanish), Ministerio de Asuntos Exteriores y de Cooperación de España.
- Spanish Ministry's Plan for Africa 2009-2012 (http://www.maec.es/es/Home/Paginas/planafrica2009_12.aspx) - (Spanish), Ministerio de Asuntos Exteriores y de Cooperación de España.
- "Guia Pais: Guinea Ecuatorial" (http://www.icex.es/staticFiles/GuineaEcuatorial_6814_.pdf) - (Spanish), by The Office for Economic and Commercial Affairs, Embassy of Spain in Lagos, Nigeria, March 2004
- History of Equatorial Guinea (http://www.pbs.org/wnet/wideangle/episodes/once-upon-a-coup/history-of-equatorial-guinea/5475/) PBS WIDE ANGLE interactive timeline

- Once Upon a Coup (http://www.pbs.org/wnet/wideangle/episodes/once-upon-a-coup/introduction/5309/) PBS WIDE ANGLE documentary about the 2004 coup attempt

bjn:Guinea Katulistiwa gag:Ekvatorial Gvineya mrj:Экваториаль Гвиней

Economy of Equatorial Guinea

Economy of Equatorial Guinea	
Rank	108th
Currency	Communaute Financiere Africaine franc (XAF)
Fiscal year	calendar year
Trade organizations	UDEAC, ECCAS
Statistics	
GDP	$10.41 billion (2009)[1]
GDP growth	18.6% (2005)
GDP per capita	$15,401 (2009)[1]
GDP by sector	*agriculture*: 2.8%, *industry*: 92.6%, *services*: 4.5% (2006 est.)
Inflation (CPI)	5.2% (2006)
Unemployment	30% (1998)
Main industries	petroleum, fishing, sawmilling, natural gas
Ease of Doing Business Rank	164th[2]
External	
Exports	$8.961 billion (2006)
Main export partners	US 24.6%, China 21.8%, Spain 10.8%, Canada 7.3%, Taiwan 7.2%, Portugal 5.5%, Netherlands 5.2%, Brazil 4.6%, France 4% (2005)
Imports	$2.543 billion f.o.b. (2006)
Main import partners	US 24.5%, Italy 20.6%, France 12.1%, Spain 10.8%, Cote d'Ivoire 8.6%, UK 6.9% (2005)
Public finances	
Public debt	*external*: $289 million (2006)
Revenues	$2.752 billion (2006)
Expenses	$1.424 billion (2006)
Economic aid	none
Main data source: CIA World Fact Book [3] *All values, unless otherwise stated, are in US dollars*	

Equatorial Guinea is a small country at the west coast of Central Africa. Despite a per capita GDP (PPP) of more than US$30,000 [4] (CIA Factbook $50,200[5]) which is as of 2008 the twentieth highest in the world, Equatorial Guinea ranks 121st out of 177 countries on the United Nations Human Development Index.

Economy overview

Pre-independence Equatorial Guinea counted on cocoa production for hard currency earnings. In 1959 it had the highest per capita income of Africa.

The discovery of large oil reserves in 1996 and its subsequent exploitation have contributed to a dramatic increase in government revenue. As of 2004,[6] Equatorial Guinea is the third-largest oil producer in Sub-Saharan Africa. Its oil production has risen to 360000 barrels per day (57000 m^3/d), up from 220000 barrels per day (35000 m^3/d) only two years earlier.

Forestry, farming, and fishing are also major components of GDP. Subsistence farming predominates. Although pre-independence Equatorial Guinea counted on cocoa production for hard currency earnings, the neglect of the rural economy under successive regimes has diminished potential for agriculture-led growth. However, the government has stated its intention to reinvest some oil revenue into agriculture. A number of aid programs sponsored by the World Bank and the IMF have been cut off since 1993 because of corruption and mismanagement. No longer eligible for concessional financing because of large oil revenues, the government has been unsuccessfully trying to agree on a "shadow" fiscal management program with the World Bank and IMF. Businesses, for the most part, are owned by government officials and their family members. Undeveloped natural resources include titanium, iron ore, manganese, uranium, and alluvial gold (Mining in Equatorial Guinea). Growth remained strong in 2005 and 2006, led by oil.

2004 US senate report

A Senate report [7] in 2004 found that Riggs Bank helped top officials of Equatorial Guinea steal hundreds of millions of dollars in oil revenues. The Senate Permanent Subcommittee on Investigations probed the corruption case involving Equatorial Guinea and its oil revenues for much of the past year.

The panel's report found that Riggs Bank helped government leaders in Equatorial Guinea siphon oil revenues to accounts set up for them in Washington. Equatorial Guinea has been cited by the U.S. State Department for human rights abuses, corruption, and diversion of oil revenues to government officials.

Beginning in 1995 until 2004, Riggs oversaw as many as 60 accounts containing as much as $700 million, making Equatorial Guinea its largest single customer. Some were government accounts, while others were the private accounts of President Teodoro Obiang Nguema, other government officials, and their families.

The Senate report says millions of dollars in the government accounts, which should have gone to help impoverished Equatorial Guineans, were instead funneled to off-shore tax shelters, with help from Riggs officials.

At a hearing on the matter at which current and former bank officials appeared, Senator Carl Levin, a Michigan Democrat, was clearly outraged. "Somehow there has to be a conscience here. Aren't you troubled?," he said.

The report describes one incident in which the bank manager of the country's accounts, Simon Kareri, brought a 27 kilogram suitcase with $3 million in plastic-wrapped cash to Riggs's Dupont Circle branch to make a deposit into President Obiang's account. Mr. Kareri, who was fired in January, refused to testify at the hearing," he said. "Mr. Chairman, there is nothing I would like to do more than answer your questions today. However, I must heed the advice of my counsel and invoke my fifth amendment rights under the Constitution and refuse to answer the question," he said.

Senators were angered that bank officials never reported any suspicious financial transactions involving Equatorial Guinea.

Lawrence Hebert, president and chief executive officer of Riggs Bank, blamed a lack of an internal system to monitor and identify such suspicious activity. It is an assertion that Senator Levin found ridiculous.

"First of all Mr. Hebert, you do not need a computer system to realize suspicious activity when you have sixty pounds of cash that are being walked into the door with a suitcase," he said.

Mr. Levin criticized bank regulators for not doing enough in their oversight responsibility.

Riggs was fined a record $25 million by federal banking regulators for allegedly failing to report suspicious transactions made to the Equatorial Guinea accounts, but that didn't happen until May 2004.

Senator Levin also took aim at oil companies that are doing business in Equatorial Guinea, many of which have secret contracts with firms that have ties to President Obiang.

The London-based organization Global Witness notes that the Senate committee report found that oil companies made payments into the personal accounts of Equatorial Guinean officials that were used for land purchases, office leases, and even education for the children of the country's leaders.

With oil money stashed away in Riggs Bank for the ruling elite of Equatorial Guinea, argues Sarah Wykes of Global Witness, oil companies cannot make the case that they are a force for positive change in the country.

"Equatorial Guinea is now the third largest producer of oil in sub-Saharan Africa," she said. "It has been called the 'Kuwait of Africa' But it is clear that since oil came on stream, the human development indicators of the country have actually gone backward, so we can say the oil money is not contributing to development at all," she said.

Equatorial Guinea responded with an 82-point report claiming that the Senate had been "duped" by "pressure groups",[8] naming specifically a $40,000 contract between a lobbying group and Severo Moto Nsa, an EquatoGuinean living in self-imposed exile who uses his website to make wild accusations against Obiang.[9] The rebuttal also claims that the payments to government members were disclosed in the government accounts and were not illegal.

In greater depth

Oil and gas exports have increased substantially (in 2003 Equatorial Guinea was ranked third among Sub-Sahara African producers behind Nigeria and Angola) and will drive the economy for years to come. The GDP increased by 105.2% in 1997, and real GDP growth reached 23% in 1999, and initial estimates suggested growth of about 15% in 2001, according to IMF 2001 forecast. Per capita income grew from about $1,000 in 1998 to about $2,000 in 2000. The energy export sector is responsible for this rapid growth. Oil production has increased from 81000 barrels per day (12900 m^3/d) to 210000 barrels per day (33000 m^3/d) between 1998 and early 2001. There is ongoing additional development of existing commercially viable oil and gas deposits as well as new exploration in other offshore concessions.

Equatorial Guinea has other largely unexploited human and natural resources, including a tropical climate, fertile soils, rich expanses of water, deepwater ports, and an untapped, if unskilled, source of labor. Following independence in 1968, the country suffered under a repressive dictatorship for 11 years, which devastated the economy. The agricultural sector, which historically was known for cocoa of the highest quality, has never fully recovered. In 1969 Equatorial Guinea produced 36,161 tons of highly bid cocoa, but production dropped to 4,800 tons in 2000. Coffee production also dropped sharply during this period to bounce back to 100,000 metric tons in 2000. Timber is the main source of foreign exchange after oil, accounting for about 12.4% of total export earnings in 1996-99. Timber production increased steadily during the 1990s; wood exports reached a record 789,000 cubic meters in 1999 as demand in Asia (mainly China) gathered pace after the 1998 economic crisis. Most of the production (mainly Okoume) goes to exports, and only 3% is processed locally. Environmentalists fear that exploitation at this level is unsustainable and point out to the permanent damage already inflicted on the forestry reserves on Bioko.

Consumer price inflation has declined from the 38.8% experienced in 1994 following the CFA franc devaluation, to 7.8% in 1998, and 1.0% in 1999, according to BEAC data. Consumer prices rose about 6% in 2000, according to initial estimates, and there was anecdotal evidence that price inflation was accelerating in 2001.

Equatorial Guinea's policies, as defined by law, comprise an open investment regime. Qualitative restrictions on imports, non-tariff protection, and many import licensing requirements were lifted when in 1992 the government adopted a public investment program endorsed by the World Bank. The Government of the Republic of Equatorial Guinea has sold some state enterprises. It is attempting to create a more favourable investment climate, and its investment code contains numerous incentives for job creation, training, promotion of non-traditional exports, support of development projects and indigenous capital participation, freedom for repatriation of profits, exemption from certain taxes and capital, and other benefits. Trade regulations have been further liberalized since implementation in 1994 of the ICN turnover tax, in conformity with Central African tax and custom reform codes. The reform included elimination of quota restrictions and reductions in the range and amounts of tariffs. The CEMAC countries agreed to replace the ICN with a value added tax (VAT) in 1999.

While business laws promote a liberalized economy, the business climate remains difficult. Application of the laws remains selective. Corruption among officials is widespread, and many business deals are concluded under non-transparent circumstances.

There is little industry in the country, and the local market for industrial products is small. The government seeks to expand the role of free enterprise and to promote foreign investment but has had little success in creating an atmosphere conducive to investor interest.

The Equato-Guinean budget has grown enormously in the past 3 years as royalties and taxes on foreign company oil and gas production have provided new resources to a once poor government. The 2001 budget foresaw revenues of about 154 billion CFA francs (154 GCFAF) (about U.S.$200 million), up about 50% from 2000 levels. Oil revenues account for about two-thirds of government revenue, and VAT and trade taxes are the other large revenue sources.

Year 2001 government expenditures were planned to reach 158 billion CFA francs, up about 50% from 2000 levels. New investment projects represented about 40% of the budget, and personnel and internal and external debt payments represented about one-third of planned expenditures.

The Equato-Guinean Government has undertaken a number of reforms since 1991 to reduce its predominant role in the economy and promote private sector development. Its role is a diminishing one, although many government interactions with the private sector are at times capricious. Beginning in early 1997, the government initiated efforts to attract significant private sector involvement through a Corporate Council on Africa visit and numerous ministerial efforts. In 1998, the government privatized distribution of petroleum products. There are now Total and Mobil stations in the country. The government has expressed interest in privatizing the outmoded electricity utility. A French company operates cellular telephone service in cooperation with a state enterprise. The government is anxious for greater U.S. investment, and President Obiang visited the U.S. three times between 1999 and 2001 to encourage greater U.S. corporate interest. Investment in agriculture, fishing, livestock, and tourism are among sectors the government would like targeted.

Equatorial Guinea's balance-of-payments situation has improved substantially since the mid-1990s because of new oil and gas production and favorable world energy prices. Exports totaled about francs CFA 915 billion in 2000 (1.25 G$US), up from CFA 437 billion (700 M$US) in 1999. Crude oil exports accounted for more than 90% of export earnings in 2000. Timber exports, by contrast, represented only about 5% of export revenues in 2000. Additional oil production coming on line in 2001, combined with methanol gas exports from the new CMS-Nomeco plant, should increase export earnings substantially.

Imports into Equatorial Guinea also are growing very quickly. Imports totaled francs CFA 380 billion (530 M$US), up from franc CFA 261 million (420 M$US) in 1999. Imports of equipment used for the oil and gas sector accounted for about three-quarters of imports in 2000. Imports of capital equipment for public investment projects reached francs CFA 30 billion in 2000, up 40% from 1999 levels.

Equatorial Guinea's foreign debt stock was approximately francs CFA 69 billion (100 M$US) in 2000, slightly less than the debt stock in 1999, according to BEAC data. Equatorial Guinea's debt service ratio fell from 20% of GDP in 1994 to only 1% in 2000. Foreign exchange reserves were increasing slightly, although they were relatively low in terms of import coverage. According to the terms of the franc CFA zone, some of these reserves are kept in an account with the French Ministry of Finance.

Equatorial Guinea in the 1980s and 1990s received foreign assistance from numerous bilateral and multilateral donors, including European countries, the United States, and the World Bank. Many of these aid programs have ceased altogether or have diminished. Spain, France, and the European Union continue to provide some project assistance, as do China and Cuba. The government also has discussed working with World Bank assistance to develop government administrative capacity.

Equatorial Guinea operated under an IMF-negotiated Enhanced Structural Adjustment Facility (ESAF) until 1996. Since then, there have been no formal agreements or arrangements. The International monetary Fund held Article IV consultations (periodic country evaluations) in 1996, 1997, and in August 1999. After the 1999 consultations, IMF directors stressed the need for Equatorial Guinea to establish greater fiscal discipline, accountability, and more transparent management of public sector resources, especially energy sector revenue. IMF officials also have emphasized the need for economic data. In 1999, the Equato-Guinean Government began attempting to meet IMF-imposed requirements, maintaining contact with IMF and the World Bank representatives. However, the new found oil wealth allowed the government to avoid improving fiscal discipline, transparency and accountability.

Infrastructure

Infrastructure is generally old and in poor condition. Surface transport is extremely limited at present, with little more than 700 kilometres of paved roads. The African Development Bank is helping to improve the paved roads from Malabo to Luba and Riaba; the Chinese are undertaking a project to link Mongomo to Bata on the mainland, and the European Union is financing an inter-states road network linking Equatorial Guinea to Cameroon and Gabon. Road maintenance is often inadequate.

Electricity is available in Equatorial Guinea's larger towns thanks to three small overworked hydropower facilities and a number of aged generators. In 1999, national production was about 13 MWh. In Malabo, the American company, CMS-Nomeco, built a 10 megawatt electricity plant financed by the government, which came in line in mid-2000, and plans to double capacity are advancing. This plant provides improved service to the capital, although there are still occasional outages. On the mainland the largest city, Bata, still has regular blackouts.

Water is only available in the major towns and is not always reliable because of poor maintenance and mismanagement. Some villages and rural areas are equipped with generators and water pumps, usually owned by private individuals.

Parastatal Getesa, a joint venture with a minority ownership stake held by a French subsidiary of France Telecom, provides telephone service in the major cities. The regular system is overextended, but France Telecom has introduced a popular GSM system, which is generally reliable in Malabo and Bata.

Equatorial Guinea has two of the deepest Atlantic seaports of the region, including the main business and commercial port city of Bata. The ports of both Malabo and Bata are severely overextended and require extensive rehabilitation and reconditioning. The British company, Incat, has an ongoing project with the government to renovate and expand Luba, the country's third-largest port which is located on Bioko Island. The government hopes Luba will become a major transportation hub for offshore oil and gas companies operating in the Gulf of Guinea. Luba is located some 50 kilometres from Malabo and had been virtually inactive except for minor fishing activities and occasional use to ease congestion in Malabo. A new jetty is also being built at km 5 on the way from Malabo to the airport. It is a project mainly supposed to service the oil industry, but can also relieve the congested Malabo Port due to its closeness. The Oil Jetty at km 5 was supposed to open the end of March 2003. Riaba is the other port of any scale on Bioko but is less active. The continental ports of Mbini and Cogo have deteriorated as well and are now used primarily for timber activities.

There are both air and sea connections between the two cities of Malabo and Bata. A few aging Soviet-built aircraft operated by several small carriers, one state-owned, and the others private, constitute the national aircraft fleet. The runway at Malabo (3,200 m) is equipped with lights and can service equipment similar to DC 10s and Cl3Os. The one at Bata (2,400 m) does not operate at night but can accommodate aircraft as large as B737s. Their primary users are the national airline (EGA) and a private company (GEASA). Two minor airstrips (800 m) are located at Mongomo and Annobon. There are international connections out of Malabo to Madrid and Zurich in Europe and to Cotonou, Douala and Libreville in West Africa.

Energy developments

After a slow start, Equatorial Guinea has recently emerged as a major oil producer in the Gulf of Guinea, one of the most promising hydrocarbon regions in the world. The main oil fields, Zafiro and Alba, are both located offshore of Bioko island. In 1999 oil production was about five times its 1996 level; Zafiro Field, operated by Exxon Mobil and Ocean Energy, produced about 100000 barrels per day (16000 m^3/d), and CMS Nomeco extracted approximately 6700 barrels per day (1070 m^3/d). Aggressive field development and promising exploration activities may raise production to nearly 300000 barrels per day (48000 m^3/d) within 2–3 years (slightly above the current estimated production of Gabon) according to the official sources.

In 1995 Mobil (now Exxon Mobil) discovered the large Zafirobbbb field, with estimated reserves of 400000000 barrels (m^3). Production began in 1996. The company announced a 3-year U.S.$1bn rapid development program to boost output to 130000 barrels per day (21000 m^3/d) by early 2001. Progress was delayed due to a contractual dispute with the government and by unexpectedly difficult geology. The difference with the government was eventually resolved.

In 1998 a more liberal regulatory and profit-sharing arrangement for hydrocarbon exploration and production activities was introduced. It revised and updated the production-sharing contract, which, until then, had favoured Western operators heavily. As a result domestic oil receipt rose from 13% to 20% of oil export revenue. However, the government's share remains relatively poor by international standards.

In 1997 CMS Nomeco moved to expand its operation with a U.S.$300m methanol plant. The plant entered production in 2000 and help boost natural gas condensate output from Alba field.

In August 1999 the government closed bidding on a new petroleum licensing round for 53 unexplored deepwater blocks and seven shallow water blocks. The response was small due to combination of factors, including falling oil prices, restructuring within the oil industry, and uncertainty over and undemarcated maritime border with Nigeria (which was not resolved until 2000).

In late 1999 Triton Energy, a U.S. independent, discovered La Ceiba in block G in an entirely new area offshore the mainland of the country. Triton expects a U.S.$200m development program to enable La Ceiba and associated fields to produce 100000 barrels per day (16000 m^3/d) by late 2001, despite disappointments and technical problems at the beginning of the year.

With an upturn in oil prices, exploration intensified in 2000. In April 2000 U.S.-based Vanco Energy signed a production-sharing contract for the offshore block of Corisco Deep. In May 2000, Chevron was granted block L, offshore Rio Muni, and a further three production-sharing contracts (for blocks J, I, and H) were signed with Atlas Petroleum, a Nigerian company.

In early 2001 the government announced plans to establish a national oil company, to allow Equatorial Guinea to take a greater stake in the sector and to facilitate the more rapid transfer of skills. However, critics fear that such a company may become a vehicle for opaque accounting and inertia of the sort that has hindered development in neighbouring countries including Angola, Cameroon, and Nigeria.

Since 2001 the government has created GEPetrol, a national oil company; and Sonagas, a national natural gas company. The company EG LNG has been created to construct and operate the Bioko Island LNG plant and terminal.

Animal Husbandry

Cattle and poultry production is rapidly reaching its pre-independence levels of self-sufficiency with the financial help of the African Development Bank. However, production of domesticated animals is hindered by the presence of trypanosomiasis and other tropical deterrents. In 2005 there were 37,600 sheep, 9,000 goats, 6,100 hogs, and 5,000 cattle.

Fishing

The fishing industry gained strength through the 1980s and is now almost entirely modernized; a tuna processing plant went into operation in 1990. Annobón subsists almost entirely on fishing and retains its traditional preeminence in off shore whaling and turtle gathering. Bioko is also a major fishing center, the chief catches being perch, tuna, mackerel, cod, pike, shark, and crayfish. The country's own catch was about 3,500 tons in 2003.

Forestry

Timber from Río Muni is Equatorial Guinea's leading export. Forests cover over 62 percent of the land area. The Río Muni area on the mainland produces okoumé and akoga from rain forests of considerable age. Even though the government has given permission to foreign firms, exploitation is difficult due to infrastructural problems. The government enacted a new forestry action plan in 1990 in an effort to strengthen the sector's development. In 2004, roundwood production was estimated at 811,000 cubic metres. In 2004, exports of forest products amounted to $97 million Child rate- 4

$24 billion (2009 est.)

Investment (gross fixed)

46.3% (2005 est.)

Industries

Petroleum, fishing, sawmilling, natural gas

Industrial production growth rate

30% (2002 est.)

Electricity - production

29.43 GWh (2005)

Electricity - consumption

27.37 GWh (2005)

Agriculture - products

Coffee, cocoa, rice, yams, cassava (tapioca), bananas, palm oil nuts; livestock; timber

Exchange rates

Communauté financière africaine francs (CFAF) per US$1 - 480.56 (2005), 528.29 (2004), 581.2 (2003), 696.99 (2002), 733.04 (2001)

References

[1] World Development Indicators database (http://data.worldbank.org/data-catalog#Tables), World Bank, accessed on July 6, 2010.
[2] "Doing Business in Equatorial Guinea 2010" (http://www.doingbusiness.org/ExploreEconomies/?economyid=63). World Bank. . Retrieved 2010-08-20.
[3] https://www.cia.gov/library/publications/the-world-factbook/geos/ek.html
[4] World Economic Outlook IMF Database, September 2005 (http://imf.org/external/pubs/ft/weo/2005/02/data/dbcoutm.cfm?SD=2004&ED=2004&R1=1&R2=1&CS=3&SS=2&OS=C&DD=0&OUT=1&C=512-941-914-446-612-666-614-672-311-946-213-137-911-962-193-674-122-676-912-548-313-556-419-678-513-181-316-682-913-684-124-273-339-921-63 S=PPPPC&CMP=0&x=18&y=9IMF)
[5] CIA - The World Factbook - Equatorial Guinea (https://www.cia.gov/library/publications/the-world-factbook/geos/ek.html)
[6] Blum, Justin (2004-09-07). "U.S. Oil Firms Entwined in Equatorial Guinea Deals" (http://www.washingtonpost.com/wp-dyn/articles/A1101-2004Sep6.html). *The Washington Post*. .
[7] Friendly Dealings With Dictators (http://www.progress.org/2004/oil13.htm)
[8] http://www.guineequatoriale-info.net/info_05/us_2005052500.htm
[9] Sitio web Oficial del Gobierno de Guinea Ecuatorial en el exilio - Manuel Tomo, hermano del miembro de FDR y DECAM, Germán Tomo Mayo, ha sido apuñalado por sicarios de Obiang Nguema este mediodia en Alcorcón (Madrid) (http://www.guinea-ecuatorial.org/modules.php?name=News&file=print&sid=448)

External links

- Economy of Equatorial Guinea (http://www.dmoz.org/Regional/Africa/Equatorial_Guinea/Business_and_Economy/Economic_Development/) at the Open Directory Project
- Equatorial Guinea latest trade data on ITC Trade Map (http://www.trademap.org/open_access/Index.aspx?proceed=true&reporter=226)
- Equatorial Oil (http://www.equatorialoil.com)- Government source on economics, run by the Ministry of Mines, Industry and Energy

Gross national income

Gross national income (GNI) comprises the value of all products and services generated within a country in one year (i.e., its gross domestic product), together with its net income received from other countries (notably interest and dividends).[1]

The GNI consists of: the personal consumption expenditures, the gross private investment, the government consumption expenditures, the net income from assets abroad (net income receipts), and the gross exports of goods and services, after deducting two components: the gross imports of goods and services, and the indirect business taxes. The GNI is similar to the gross national product (GNP), except that in measuring the GNP one does not deduct the indirect business taxes.

GNI versus GDP

For example, the profits of a US-owned company operating in the UK will only count towards US GNI and UK GDP. If a country becomes heavily indebted, and pays large amounts of interest to service this debt, this will be reflected in a decreased GNI but not a decreased GDP. If a country sells off its resources to entities outside their country this will also be reflected over time in decreased GNI, but not decreased GDP. Therefore, the GDP appears more attractive for countries with increasing national debt and decreasing assets.

GNI versus GNP

GNP is a concept that goes hand in hand with GNI, GDP, and NNI. In contrast to the GNI, the GNP does not account for the balance of cross-country income, such as interest and dividends. In contrast to the GDP, the GNP account for the values of products and services based on citizenship of the owners rather than the territory of the activity.

List of countries by GNI (nominal, Atlas method) millions of US$[2] (Top 10)

Rank	2010		2009		2008	
1	United States	14,600,828	United States	14,223,686	United States	14,506,142
2	China	5,700,018	China	4,857,623	Japan	4,853,005
3	Japan	5,369,116	Japan	4,785,450	China	4,042,883
4	Germany	3,537,180	Germany	3,473,814	Germany	3,504,510
5	France	2,749,821	France	2,750,418	United Kingdom	2,799,960
6	United Kingdom	2,399,292	United Kingdom	2,538,578	France	2,700,770
7	Italy	2,125,845	Italy	2,114,668	Italy	2,115,482
8	Brazil	1,830,392	Brazil	1,563,126	Spain	1,449,186
9	India	1,566,636	Spain	1,472,046	Canada	1,446,669
10	Spain	1,462,894	India	1,405,064	Brazil	1,433,699

References

[1] Lequiller, François; Derek Blades (2006). *Understanding National Accounts* (http://books.google.co.uk/books?id=pXpJL6f8b3wC&printsec=frontcover&dq="Understanding+National+Accounts"&source=bl&ots=_6_lHq-McY&sig=YqWljozkylpi4IFspFnjGwPicPw&hl=en&ei=4g7GTLT7OJCQjAfXiZ11&sa=X&oi=book_result&ct=result&resnum=5&ved=0CCMQ6AEwBA#v=onepage&q="To convert GDP into GNI"&f=false). OECD. pp. 18–19. ISBN 978-92-64-02566-0. .

[2] (http://data.worldbank.org/indicator/NY.GNP.ATLS.CD)

See also

- Net national income
- Measures of national income and output
- Gross national income in the European Union
- List of countries by GNI per capita
- Low-Income Countries Under Stress (LICUS) (World Bank program)
- National average salary

Hydrocarbon exploration

Hydrocarbon exploration (or **oil and gas exploration**) is the search by petroleum geologists and geophysicists for hydrocarbon deposits beneath the Earth's surface, such as oil and natural gas. Oil and gas exploration are grouped under the science of petroleum geology.

List of components of oil drilling rigsOnshore Drilling Rig

Exploration methods

Visible surface features such as oil seeps, natural gas seeps, pockmarks (underwater craters caused by escaping gas) provide basic evidence of hydrocarbon generation (be it shallow or deep in the Earth). However, most exploration depends on highly sophisticated technology to detect and determine the extent of these deposits using exploration geophysics. Areas thought to contain hydrocarbons are initially subjected to a gravity survey, magnetic survey, passive seismic or regional seismic reflection surveys to detect large scale features of the sub-surface geology. Features of interest (known as *leads*) are subjected to more detailed seismic surveys which work on the principle of the time it takes for reflected sound waves to travel through matter (rock) of varying densities and using the process of depth conversion to create a profile of the substructure. Finally, when a prospect has been identified and evaluated and passes the oil company's selection criteria, an exploration well is drilled in an attempt to conclusively determine the presence or absence of oil or gas.

Oil exploration is an expensive, high-risk operation. Offshore and remote area exploration is generally only undertaken by very large corporations or national governments. Typical shallow shelf oil wells (e.g. North Sea) cost USD$10 – 30 million, while deep water wells can cost up to USD$100 million plus. Hundreds of smaller companies search for onshore hydrocarbon deposits worldwide, with some wells costing as little as USD$100,000.

Elements of a petroleum prospect

A prospect is a potential trap which geologists believe may contain hydrocarbons. A significant amount of geological, structural and seismic investigation must first be completed to redefine the potential hydrocarbon drill location from a lead to a prospect. Five geological factors have to be present for a prospect to work and if any of them fail neither oil nor gas will be present.

Mud log in process, a common way to study the rock types when drilling oil wells.

- A source rock - When organic-rich rock such as oil shale or coal is subjected to high pressure and temperature over an extended period of time, hydrocarbons form.
- Migration - The hydrocarbons are expelled from source rock by three density-related mechanisms: the newly-matured hydrocarbons are less dense than their precursors, which causes overpressure; the hydrocarbons are lighter medium, and so migrate upwards due to buoyancy, and the fluids expand as further burial causes increased heating. Most hydrocarbons migrate to the surface as oil seeps, but some will get trapped.
- Trap - The hydrocarbons are buoyant and have to be trapped within a structural (e.g. Anticline, fault block) or stratigraphic trap
- Seal or cap rock - The hydrocarbon trap has to be covered by an impermeable rock known as a seal or cap-rock in order to prevent hydrocarbons escaping to the surface
- Reservoir - The hydrocarbons are contained in a reservoir rock. This is a porous sandstone or limestone. The oil collects in the pores within the rock. The reservoir must also be permeable so that the hydrocarbons will flow to surface during production.

Exploration risk

Hydrocarbon exploration is a high risk investment and risk assessment is paramount for successful exploration portfolio management. Exploration risk is a difficult concept and is usually defined by assigning confidence to the presence of five imperative geological factors, as discussed above. This confidence is based on data and/or models and is usually mapped on Common Risk Segment Maps (CRS Maps). High confidence in the presence of imperative geological factors is usually colored green and low confidence colored red [1] . Therefore these maps are also called Traffic Light Maps, while the full procedure is often referred to as Play Fairway Analysis [2] . The aim of such procedures is to force the geologist to objectively assess all different geological factors. Furthermore it results in simple maps that can be understood by non-geologists and managers to base exploration decisions on.

Terms used in petroleum evaluation

- Bright spot - On a seismic section, coda that have high amplitudes due to a formation containing hydrocarbons.
- Chance of success - An estimate of the chance of all the elements (see above) within a prospect working, described as a probability. High risk prospects have a less than 10% chance of working, medium risk prospects 10-20%, low risk prospects over 20%. Typically about 40% of wells recently drilled find commercial hydrocarbons.
- Dry hole - A formation that contains brine instead of oil.
- Flat spot - An oil-water contact on a seismic section; flat due to gravity.
- Hydrocarbon in place - amount of hydrocarbon likely to be contained in the prospect. This is calculated using the volumetric equation - GRV x N/G x Porosity x Sh x FVF
  - GRV - Gross rock volume - amount of rock in the trap above the hydrocarbon water contact
  - N/G - net/gross ratio - proportion of the GRV formed by the reservoir rock (range is 0 to 1)

  - Porosity - percentage of the net reservoir rock occupied by pores (typically 5-35%)
  - Sh - hydrocarbon saturation - some of the pore space is filled with water - this must be discounted
  - FVF - formation volume factor - oil shrinks and gas expands when brought to the surface. The FVF converts volumes at reservoir conditions (high pressure and high temperature) to storage and sale conditions
- Lead - a structure which may contain hydrocarbons
- Play - A particular combination of reservoir, seal, source and trap associated with proven hydrocarbon accumulations
- Prospect - a lead which has been fully evaluated and is ready to drill
- Recoverable hydrocarbons - amount of hydrocarbon likely to be recovered during production. This is typically 10-50% in an oil field and 50-80% in a gas field.

Licensing

Petroleum resources are typically owned by the government of the host country. In the USA most onshore (land) oil and gas rights (OGM) are owned by private individuals. Sometimes this is not the same person who owns the surface rights. In this case oil companies must negotiate terms for a lease of these rights with the individual who owns the OGM. In most nations the government issues licences to explore, develop and produce its oil and gas resources, which are typically administered by the oil ministry. There are several different types of licence. Typically oil companies operate in joint ventures to spread the risk, one of the companies in the partnership is designated the operator who actually supervises the work.

- Tax and Royalty - Companies would pay a royalty on any oil produced, together with a profits tax (which can have expenditure offset against it). In some cases there are also various bonuses and ground rents (license fees) payable to the government - for example a signature bonus payable at the start of the licence. Licences are awarded in competitive bid rounds on the basis of either the size of the work programme (number of wells, seismic etc.) or size of the signature bonus.
- Production Sharing contract (PSA) - A PSA is more complex than a Tax/Royalty system - The companies bid on the percentage of the production that the host government receives (this may be variable with the oil price), There is often also participation by the Government owned National Oil Company (NOC). There are also various bonuses to be paid. Development expenditure is offset against production revenue.
- Service contract - This is when an oil company acts as a contractor for the host government, being paid to produce the hydrocarbons.

Reserves and resources

Resources are hydrocarbons which may or may not be produced in the future. A resource number may be assigned to an undrilled prospect or an unappraised discovery. Appraisal by drilling additional delineation wells or acquiring extra seismic data will confirm the size of the field and lead to project sanction. At this point the relevant government body gives the oil company a production licence which enables the field to be developed. This is also the point at which oil reserves can be formally booked.

Definition of oil reserves

Oil reserves are primarily a measure of geological risk - of the probability of oil existing and being producible under current economic conditions using current technology. The three categories of reserves generally used are proven, probable, and possible reserves.

- **Proven reserves** - defined as oil and gas "Reasonably Certain" to be producible using current technology at current prices, with current commercial terms and government consent- also known in the industry as 1P. Some Industry specialists refer to this as P90 - i.e. having a 90% certainty of being produced.
- **Probable reserves** - defined as oil and gas "Reasonably Probable" of being produced using current or likely technology at current prices, with current commercial terms and government consent - Some Industry specialists refer to this as P50 - i.e. having a 50% certainty of being produced. - This is also known in the industry as 2P or Proven plus probable.
- **Possible reserves** - i.e. "having a chance of being developed under favourable circumstances" - Some industry specialists refer to this as P10 - i.e. having a 10% certainty of being produced. - This is also known in the industry as 3P or Proven plus probable plus possible.

Reserve booking

Oil and gas reserves are the main asset of an oil company - booking is the process by which they are added to the Balance sheet. This is done according to a set of rules developed by the Society of Petroleum Engineers (SPE). The Reserves of any company listed on the New York Stock Exchange have to be stated to the U.S. Securities and Exchange Commission. In many cases these reported reserves are audited by external geologists, although this is not a legal requirement. The U.S. Securities and Exchange Commission rejects the probability concept and prohibits companies from mentioning probable and possible reserves in their filings. Thus, official estimates of proven reserves will always be understated compared to what oil companies think actually exists. For practical purposes companies will use proven plus probable estimate (2P), and for long term planning they will be looking primarily at possible reserves.

Other countries also have their national hydrocarbon reserves authorities for example, Russia's State Commission on Mineral Reserves (GKZ), to which companies operating in these countries have to report.

See also

- Wildcatter
- Abiogenic petroleum origin
- Energy development
- Future energy development
- Hubbert peak
- Petroleum
- Renewable energy
- Giant oil and gas fields
- Upward continuation
- NORM
- Drill baby drill

References

[1] *Exploration Risk on E&P Geology* http://www.epgeology.com/general-discussion-f28/assigning-exploration-risks-t157.html
[2] *CRS Mapping and Play Fairway Analysis* http://www.ccop.or.th/projects/PPM/Case_Study_Phillipines_files/Second%20workshop/presentations%5CP12%20PhilPRA%20Play%20Analysis%20presentation%202nd%20workshop.pdf

External links

- Oil & Gas Exploration News (http://www.ogj.com/category/Online_SubCategory.cfm?p=7&Section=OnlineArticles&cat=ExplD)
- Interactive Oil Well Map (http://www.mississippioiljournal.com/map/) Mississippi Oil Journal
- Overview of oil and gas exploration (http://www.petrostrategies.org/Learning_Center/exploration.htm)
- Oilfield Glossary (http://www.glossary.oilfield.slb.com/)
- Exploration Geology Forums (http://www.epgeology.com/)

Spanish Empire

Spanish Empire

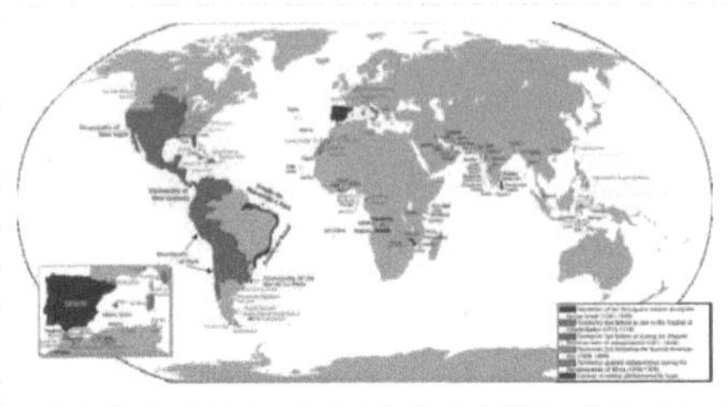

The areas of the world that at one time were territories of the Spanish Empire.

Portuguese territories during the Iberian Union (1581–1640).

Territories held before the Treaties of Utrecht–Baden (1713–1714).

Territories held before the Spanish American wars of independence (1811–1828).

Territories held before the Spanish-American War (1898–1899).

Territories granted independence during the Decolonization of Africa (1956–1976).

Current territories administered by Spain.

The **Spanish Empire** (Spanish: *Imperio Español*) consisted of the territories and colonies administered directly by Spain in Europe, the Americas, Africa, Asia and Oceania. It originated during the Age of Exploration and was therefore one of the first global empires. At the time of Habsburgs, Spain reached the peak of its world power.[1] It lasted from the 15th century through — in the case of its African holdings — the latter portion of the 20th century.

After the War of the Castilian Succession, Spain had emerged with a personally unified monarchy, by the marriage of the Catholic Monarchs or *los Reyes Catolicos*: the Queen of Castile and the King of Aragon. Rule was separate but internal and foreign policy was coordinated. In 1492, the Spanish monarchs completed the *Reconquista* with the incorporation of Granada to the Kingdom of Castile. That same year Christopher Columbus commanded the first Spanish exploratory voyage west across the Atlantic Ocean, leading to the Discovery of America and Europe's eventual colonial engagement in the New World. The Americas thereby became the focus of Spanish exploration and colonization.

In the 16th century, Spain settled the Greater Antilles in the Caribbean, and took over large areas on mainland North and South America overrunning the Aztecs and Incas. The Spanish expedition of world circumnavigation started by Ferdinand Magellan in 1519, and completed by Juan Sebastian Elcano in 1522, achieved what Columbus had longed for, a westward route to Asia and the sought-after Spice Islands. In 1565 navigator Miguel Lopez de Legazpi arrived in Guam and the Philippine Islands establishing the Spanish East Indies. In addition to the overseas empire in America and Oceania, the Spanish Monarchy controlled several European territories (the Low Countries, the greater part of Italy, and some parts of modern France and Germany), and a number of coastal strongholds in Africa. By the 17th century, Spain controlled an empire on a scale and world distribution that had never been approached by its predecessors.[2]

Spain's European possessions were given up at the conclusion of the War of the Spanish Succession in 1713, but Spain retained its vast overseas empire. In 1741, a massive victory over Britain at the Battle of Cartagena de Indias in modern day Colombia prolonged Spain's hegemony in the Americas until the 19th century. During the late 18th century, Spanish expeditions to the Pacific Northwest reached Canada and Alaska, resulting in a settlement on Vancouver Island and the discovery of several archipelagos and glaciers.

The French occupation of Spain in 1808 under Napoleon cut off its American colonies temporarily, and a number of independence movements between 1810 and 1825 resulted in a chain of newly independent Spanish American republics in South and Central America. The remainder of Spain's then-four hundred year empire, namely Cuba, Puerto Rico, and the Spanish East Indies, continued under Spanish control until the end of the 19th century, when most of these territories were annexed by the United States after the Spanish-American War. Spain sold its remaining Pacific islands to Germany in 1899. Therefore, at the turn of the 20th century, Spain only held territories in Africa, namely Spanish Guinea, Spanish Sahara and Spanish Morocco, obtained during the Scramble for Africa, but they were relinquished from the mid 20th century due to the Decolonization of Africa.

Definition

The land of the Iberian peninsula was commonly called Hispania since Roman times and during the Visigothic Kingdom. The process of the Reconquista produced the emergence of four Christian realms: Castile, Aragon, Navarre and Portugal.

The dynastic union between the Crown of Castile (which included the kingdom of Navarre) and the Crown of Aragon,[3] in Catholic Monarchs's time, initiated a political system in force until the beginning of the eighteenth century labelled as Hispanic monarchy: the Spanish sovereign acted as monarch in a unitary manner[4] over all his territories through a polisynodial system of Councils, but his power as king or lord varied from one territory to another one, since each territory retained its own particular administration and juridical configuration. The unity did not mean uniformity. According to this political configuration, independently of the denominations [5] given to the "dynastic union" [6] [7] between 1580–1640, the scholars argue that the Portuguese Empire kept its own administration and jurisdiction over its territory as the other kingdoms and realms ruled by the Spanish Habsburgs.[8] Nevetheless, whereas some historians assert that at that time, Portugal was a kingdom which formed part of the Spanish Monarchy;[9] [10] [11] [12] [13] others draw a clear distinction between the Portuguese Empire and the Spanish Empire.[14] [15]

The Spanish Empire includes Spain's *overseas* colonies in the Americas, Asia, Oceania and Africa, but some disputes exist as to which European territories are to be counted. For instance, normally the Habsburg Netherlands are included as they were part of the possessions of the King of Spain, governed by Spanish officials, and defended by Spanish troops. However, authors like the British historian Henry Kamen contend that these territories were not fully integrated into a Spanish state and instead formed part of the wider Habsburg possessions. Some historians use "Habsburg" and "Spanish" almost interchangeably when referring to the dynastic inheritance of Charles V or Philip II.

Origins

During the 15th century, Castile and Portugal fell within territorial and commercial rivalry in the Atlantic western zone. Portugal obtained several Papal bulls which acknowledged the Portuguese control over the discovered territories, but Castile also obtained from the Pope, the safeguard of its rights to the Canary Islands with the bulls *Romani Pontifex* dated on 6 November 1436 and *Dominatur Dominus* dated on 30 April 1437.[16] The Conquest of the Canary Islands, inhabited by Guanche people, began under the reign of Henry III of Castile in 1402, authorizing under feudal agreement to Norman noblemen Jean de Béthencourt. The conquest only finished when the armies of the Crown of Castille won, in long and bloody wars, the islands of Gran Canaria (1478–1483), La Palma (1492–1493) and Tenerife (1494–1496).

The proclamation of Isabella I of Castile jointly with her husband Ferdinand of Aragon, at that time king of Sicily, as kings of Castile drove to a war with Portugal solved with the Treaty of Alcáçovas (4 September 1479). Portuguese gave up their claims to the Canary Islands but were secured their sovereignty over Madeira, Azores and the Cape Verde islands, and over the trade of Guinea,[17] from the south of the Cape Bojador,[18] and the right to conquer the Kingdom of Fez.[19] [20] This treaty was confirmed in 1481 by the Pope Sixtus IV, in the papal bull *Æterni regis* (dated on 21 June 1481).[21]

Seven months before the treaty of Alcaçovas, the king John II of Aragon passed away and his son Ferdinand II of Aragon inherited the thrones of the Crown of Aragon; therefore, it was created with Castile a personal union, each with their own administrations, but ruled by to common monarchy.[22]

Columbus and the Catholic Monarchs (The return of Columbus to Spain)

After a war of 10 years, the Granada War, in 1492, the *Reyes Católicos* drove out the last Moorish king of Granada. After their victory, the Catholic monarchs negotiated with Christopher Columbus, a Genoese sailor attempting to reach Cipangu by sailing west. Castile was already engaged in a race of exploration with Portugal to reach the Far East by sea when Columbus made his bold proposal to Isabella. In the Capitulations of Santa Fe, dated on April 17, 1492, Christopher Columbus obtained from the Catholic Monarchs the appointment as viceroy and governor in the lands *already discovered*[23] and that of he might discover thenceforth;[24] [25] thereby, it was the first document to establish an administrative organization in the Indies.[26] Columbus' discoveries inaugurated the Spanish colonization of the Americas.

Spain's claim to these lands was solidified by the *Inter caetera* papal bull dated on 4 May 1493, and *Dudum siquidem* on 26 September 1493. Since the Portuguese wanted to keep the line of demarcation of Alcaçovas running east and west along the latitude south of Cape Bojador, a compromise was worked out which was incorporated by the Treaty of Tordesillas dated on 7 June 1494, in which the globe was divided into two hemispheres between Spanish and Portuguese claims. These actions gave Spain exclusive rights to establish colonies in all of the New World from Alaska to Cape Horn (except Brazil), as well as the easternmost parts of Asia. The treaty of Tordesillas was confirmed by the Pope Julius II in the bull *Ea quae pro bono pacis* on 24 January 1506.[27] Spain's expansion and colonization was driven by economic influences, a yearning to improve national prestige, and a desire to spread Catholicism into the New World.

On the other hand, the treaty of Tordesillas[28] and the treaty of Cintra (18 September 1509)[29] established the limits of the Kingdom of Fez for Portugal, and outside of these limits the Castilian expansion was allowed, beginning with the conquest of Melilla in 1497.

Struggles for Italy

The Catholic Monarchs had developed a marriage politics aiming at isolating their traditional enemy: France. The Monarchs' children got married with the heirs of Portugal, England and the House of Habsburg. Following the same strategy, the Catholic Monarchs decided to support the Aragonese house of Naples against Charles VIII of France in the Italian Wars from 1494. As King of Aragon, Ferdinand had been involved in the struggle against France and Venice for control of Italy; these conflicts became the center of Ferdinand's foreign policy as king. In these battles, which established the supremacy of the Spanish Tercios in the European battlefields, the forces of the kings of Spain acquired a reputation for invincibility that would last until the mid-17th century.

The death of French general Gaston de Foix at the Battle of Ravenna (1512).

After the death of Queen Isabella, Ferdinand as Spain's sole monarch adopted a more aggressive policy than he had as Isabella's husband, enlarging Spain's sphere of influence in Italy and against France. Ferdinand's first investment of Spanish forces came in the War of the League of Cambrai against Venice, where the Spanish soldiers distinguished themselves on the field alongside their French allies at the Battle of Agnadello (1509). Only a year later, Ferdinand became part of the Holy League against France, seeing a chance at taking both Milan—to which he held a dynastic claim—and Navarre. The war was less of a success than that against Venice, and in 1516, France agreed to a truce that left Milan in its control and recognized Spanish control of Upper Navarre.

First settlements in America

The Capitulations of Santa Fe granted excessive power to Columbus, and the Catholic Monarchs reacted when Colon discovered mainland in 1498.[30] The news were learnt in May 1499, and taking advantage of a revolt against Columbus in *La Española*, appointed to Francisco de Bobadilla as as governor of the Indies with civil and criminal jurisdiction over the lands discovered by Columbus, but he was soon replaced by Nicolás de Ovando in September 1501.[31] Thenceforth the Crown could authorize to individuals voyages to discover territories in the Indies with previous license,[32] and since 1503, the monopoly of the Crown was assured by the *Casa de Contratación* at Sevile. But the successors of Columbus litigated against the Crown until 1536[33] [34] for the fulfillment of the Capitulations of Santa Fe in the *pleitos colombinos*.

Following the settlement of Hispaniola which was successful towards the end of the 15th century, the colonists began searching elsewhere to begin new settlements. Those from the less prosperous Hispaniola were eager to search for new success in a new settlement. From there Juan Ponce de León conquered Puerto Rico and Diego Velázquez took Cuba.

In 1508, the Board of Navigators met in Burgos concurred the need to colonize the mainland, that which was entrusted to Alonso de Ojeda and Diego de Nicuesa as governors subordinated to the governor of Hispaniola,[35] who was the newly appointed Diego Columbus,[36] [37] with the same legal authority that Ovando.[38] The first settlement on the mainland was Darién in Panama, settled by Vasco Núñez de Balboa in 1512. In 1513, Balboa crossed the Isthmus of Panama, and led the first European expedition to see the Pacific Ocean from the West coast of the New World. In an action with enduring historical import, Balboa claimed the Pacific Ocean and all the lands adjoining it for the Spanish Crown.

The judgment of Seville of May 1511 recognized the viceregal title to Diego Columbus but limited to Hispaniola and to the islands discovered by his father, Christopher Columbus,[39] nevertheless his power was limited by royal officers and magistrates[40] [41] constituting a dual regime of government.[42] Therefore, the king Ferdinand II of Aragon as regent of his daughter the queen Joanna separated the territories of mainland, designated as Castilla de Oro,[43] from the viceroy of Hispaniola, establishing as General Lieutenant to Pedrarias Dávila in 1513[44] with functions similar to those of a viceroy, remaining Balboa subordinated as governing of Panama and Coiba[45] [46] on the Pacific Coast,[47] and that after his death returned to *Castilla de Oro*. The territory of *Castilla de Oro* did not include either Veragua (which was comprised approximately between the river Chagres[48] and cape Gracias a Dios[49]), due to this territory was subject to a lawsuit between the Crown and Diego Columbus, or the region farther north, towards the Yucatan, explored by Yáñez Pinzón and Solís in 1508-1509,[50] due to its remoteness.[51] The conflicts of the viceroy Columbus with the royal officers and with the *Audiencia*, created in 1511,[52] [53] caused his return to the Peninsula in 1515.

Campaigns in Africa

After the conquest of Melilla in 1497, the Spanish expansionist policy in North Africa was developed during the regency of Ferdinand the Catholic in Castile, stimulated by the Cisneros, once the *Reconquista* in the Iberian Peninsula was finished. That way, several towns and outposts in the North African coast were conquered and occupied by Castile: Mazalquivir (1505), Peñón de Vélez de la Gomera (1508), Oran (1509), Algiers (1510), Bugia (1510), and Tripoli (1511). In the Atlantic coast, Spain took possession of the outpost of Santa Cruz de Mar Pequeña (1476) with support from the Canary Islands, and it was retained until 1525 with the consent of the treaty of Cintra (1509).

The Spanish Habsburgs and the Golden Age (1516–1643)

The 16th and 17th centuries are sometimes called "the Golden Age of Spain" (in Spanish, *Siglo de Oro*). As a result of the marriage politics of the *Reyes Católicos*, their Habsburg grandson Charles inherited the Castilian empire in America, the Aragonese Empire in the Mediterranean (including a large portion of modern Italy), as well as the crown of the Holy Roman Empire and of the Low Countries, Franche-Comté, and Austria (this one, along with the rest of hereditary Habsburg domains was almost immediately transferred to Ferdinand, the Emperor's brother). After his defeat of the Castilian rebels (*Comuneros*) in the Castilian War of the Communities, Charles became the most powerful man in Europe, his rule stretching over an empire in Europe unrivalled in extent until the Napoleonic era. It was often said during this time that it was the empire on which the sun never set. This sprawling overseas empire of the Spanish Golden Age was controlled, not from inland Valladolid, but from Seville.

The Pillars of Hercules with the motto "Plus Ultra" as symbol of the Emperor Charles V in the Town Hall of Seville (16th century)

The Castilian Empire abroad was initially a disappointment. It did stimulate some trade and industry, but the trading opportunities encountered were limited. Matters began to change in the 1520s with the large scale extraction of silver from the rich deposits of Mexico's Guanajuato region, but it was the opening of the silver mines in Mexico's Zacatecas and Bolivia's Potosí in 1546 that became legendary. During the 16th century, Spain held the equivalent of US$1.5 trillion (1990 terms) in gold and silver received from New Spain. Ultimately, however, these imports diverted investment away from other forms of industry and contributed to inflation in Spain in the last decades of the 16th century: "I learnt a proverb here", said a French traveler in 1603: "Everything is dear in Spain except silver".[54] This situation was aggravated (though not as much as popular myth asserts) by the loss of much the commercial and artisan classes with the expulsions of the Jews (1492) and Moriscos (1609). The vast imports of silver ultimately made Spain overly dependent on foreign sources of raw materials and manufactured goods.

The wealthy preferred to invest their fortunes in public debt (*juros*), which were backed by these silver imports, rather than in production of manufactures and the improvement of agriculture. This helped perpetuate the medieval aristocratic prejudice that saw manual work as dishonorable long after this attitude had started to decline in other west European countries. The silver and gold whose circulation helped facilitate the economic and social revolutions in the Low Countries, France and England and other parts of Europe helped stifle them in Spain. The problems caused by inflation were discussed by scholars at the School of Salamanca and *arbitristas* but they had no impact on the Habsburg government.

The Habsburg dynasty spent the Castilian and American riches in wars across Europe on behalf of Habsburg interests, defaulted on their debt several times, and left Spain bankrupt several times. Tensions between the Empire and the people of Castile exploded in the popular rebellion of the Castilian War of the Communities (1520–22).

The Habsburgs' political goals were several:

- Access to the resources of the Americas (gold, silver, sugar) and products of Asia (porcelain, spices, silk)
- Undermining the power of France and containing it in its eastern borders.
- Maintaining Catholic Habsburg hegemony in Germany, defending Catholicism against the Protestant Reformation. Charles attempted to quell the Reformation at the Diet of Worms but Martin Luther refused to recant his 'heresy.' However, Charles's piety could not stop his mutinying troops from plundering the Holy See in the *Sacco di Roma*.
- Defending Europe against Islam, notably the Ottoman Empire.
- To spread religion to the unconverted souls of the new world. With conflict between Catholics and Protestants raging in Europe, the new world was an ideal place for more Catholics to be recruited.

Spanish intervention in Europe

Struggles of Charles V for Italy

With the ascent of the king Charles I in 1516 and his election as sovereign of the Holy Roman Empire in 1519, Francis I of France found himself surrounded by Habsburg territories, invaded the Spanish possessions in Italy in 1521, and inaugurated the second war of Franco-Spanish conflict. The war was a disaster for France, which suffered defeat at the Battle of Biccoca (1522), the Battle of Pavia (1525, at which Francis was captured), and the Battle of Landriano (1529) before Francis relented and abandoned Milan to Spain once more.

The Battle of Pavia (1525)

King Charles I (Charles V (Holy Roman Emperor) achieved victory at the Battle of Pavia in 1525 and surprised many Italians and Germans and elicited concerns that Charles would endeavor to gain ever greater power. Pope Clement VII switched sides and now joined forces with France and prominent Italian states against the Habsburg Emperor, in the War of the League of Cognac. In 1527, Charles grew exhausted with the pope's meddling in what he viewed as purely secular affairs, and sacked Rome itself, embarrassing the papacy sufficiently enough that Clement, and succeeding popes, were considerably more circumspect in their dealings with secular authorities. In 1533, Clement's refusal to annul the first marriage of King Henry VIII of England was a direct consequence of his unwillingness to offend the emperor and have his capital sacked for perhaps a second time. The Peace of Barcelona, signed between Charles V and the Pope in 1529, established a more cordial relationship between the two leaders. Spain was effectively named the protector of the Catholic cause and Charles was crowned as King of Italy (Lombardy) in return for Spanish intervention in overthrowing the rebellious Florentine Republic.

In 1528, the great admiral Andrea Doria allied with the Emperor to oust the French and restore Genoa's independence, opening the prospect for financial renewal: 1528 marks the first loan from Genoese banks to Charles.[55]

In 1543, the king of France Francis I announced his unprecedented alliance with the Ottoman sultan, Suleiman the Magnificent, by occupying the Spanish-controlled city of Nice in concert with Ottoman forces. Henry VIII of England, who bore a greater grudge against France than he held against the Emperor for standing in the way of his divorce, joined Charles in his invasion of France. Although the Spanish army was defeated at the Battle of Ceresole in Savoy the French were unable to seriously threaten Spanish controlled Milan, whilst suffering defeat in the north at the hands of Henry, thereby being forced to accept unfavourable terms. The Austrians, led by Charles's younger brother Ferdinand, continued to fight the Ottomans in the east. Charles went to take care of an older problem: the Schmalkaldic League.

Religious conflicts in the Holy Empire

The Schmalkaldic League had allied itself to the French, and efforts in Germany to undermine the League had been rebuffed. Francis's defeat in 1544 led to the annulment of the alliance with the Protestants, and Charles took advantage of the opportunity. He first tried the path of negotiation at the Council of Trent in 1545, but the Protestant leadership, feeling betrayed by the stance taken by the Catholics at the council, went to war, led by the Saxon elector Maurice. In response, Charles invaded Germany at the head of a mixed Dutch–Spanish army, hoping to restore the Imperial authority. The emperor personally inflicted a decisive defeat on the Protestants at the historic Battle of Mühlberg in 1547. In 1555, Charles signed the Peace of Augsburg with the Protestant states and restored stability in Germany on his principle of *cuius regio, eius religio*, a position unpopular with Spanish and Italian clergymen. Charles's involvement in Germany would establish a role for Spain as protector of the Catholic, Habsburg cause in the Holy Roman Empire; the precedent would lead, seven decades later, to involvement in the war that would decisively end Spain as Europe's leading power.

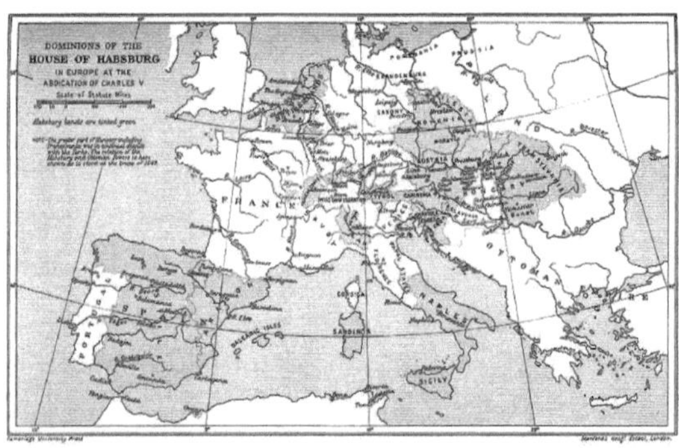

A map of the dominion of the Habsburgs following the abdication of Charles V (1556) as depicted in *The Cambridge Modern History Atlas* (1912); Habsburg lands are shaded green. From 1556 the lands in a line from the Netherlands, through to the east of France, to the south of Italy and the islands were retained by the Spanish Habsburgs.

Defeat of France

Charles V's only legitimate son, Philip II of Spain (r. 1556–98) parted the Austrian possessions with his uncle Ferdinand. Philip treated Castile as the foundation of his empire, but the population of Castile (that was about ⅓ of France's) was never great enough to provide the soldiers needed to support the Empire. When he married Mary Tudor, England was allied to Spain.

The celebrations following the Peace of Cateau-Cambrésis (1559) between Spain and France

Spain was not yet at peace, as the aggressive Henry II of France came to the throne in 1547 and immediately renewed conflict with Spain. Charles's successor, Philip II, aggressively prosecuted the war against France, crushing a French army at the Battle of St. Quentin in Picardy in 1558 and defeating Henry again at the Battle of Gravelines. The Peace of Cateau-Cambrésis, signed in 1559, permanently recognized Spanish claims in Italy. In the celebrations that followed the treaty, Henry was killed by a stray splinter from a lance. France was stricken for the next thirty years by chronic civil war and unrest (see French Wars of Religion) and removed from effectively competing with Spain and the Habsburg family in European power games. Freed from effective French opposition, Spain saw the apogee of its might and territorial reach in the period 1559–1643.

The opening for the Genoese banking consortium was the state bankruptcy of Philip II in 1557, which threw the German banking houses into chaos and ended the reign of the Fuggers as Spanish financiers.[56] The Genoese bankers provided the unwieldy Habsburg system with fluid credit and a dependably regular income. In return the less dependable shipments of American silver were rapidly transferred from Seville to Genoa, to provide capital for further ventures.

European conflicts at the time of Philip II

The time for rejoicing in Madrid was short-lived. In 1566, Calvinist-led riots in the Netherlands prompted the Duke of Alba to march into the country to restore order. In 1568, William of Orange, better known as William the Silent, led a failed attempt to drive Alva from the Netherlands. These battles are generally considered to signal the start of the Eighty Years' War that ended with the independence of the United Provinces. The Spanish, who derived a great deal of wealth from the Netherlands and particularly from the vital port of Antwerp, were committed to restoring order and maintaining their hold on the provinces. According to Luc-Normand Tellier, "It is estimated that the port of Antwerp was earning the Spanish crown seven times more revenues than the Americas."[57] In 1572, a band of rebel Dutch privateers known as the *watergeuzen* ("Sea Beggars") seized a number of Dutch coastal towns, proclaimed their support for William and denounced the Spanish leadership.

Otto van Veen: *The Relief of Leiden* (1574) after the Dutch had broken their dykes in the Eighty Years' War

For Spain, the war became an endless quagmire, sometimes literally. In 1574, the Spanish army under Luis de Requeséns was repulsed from the Siege of Leiden after the Dutch broke the dykes, thus causing extensive flooding. In 1576, faced with the bills from his 80,000-man army of occupation in the Netherlands, the cost of his fleet that had won at Lepanto, together with the growing threat of piracy in the open seas reducing his income from his American colonies, Philip was forced to accept bankruptcy. The army in the Netherlands mutinied not long after, seizing Antwerp and looting the southern Netherlands, prompting several cities in the previously peaceful southern provinces to join the rebellion. The Spanish chose to negotiate, and pacified most of the southern provinces again with the Union of Arras in 1579. In response, the Netherlands created the Union of Utrecht, as an alliance between the northern provinces, later that month. They officially deposed Philip in 1581 when they enacted the Act of Abjuration.

Under the Arras agreement the southern states of the Spanish Netherlands, today in Wallonia and the Nord-Pas-de-Calais (and Picardy) *régions* in France, expressed their loyalty to the Spanish king Philip II and recognized his Governor-General, Don Juan of Austria. In 1580, this gave King Philip the opportunity to strengthen his position when the last member of the Portuguese royal family, Cardinal Henry of Portugal, died. Philip asserted his claim to the Portuguese throne and in June sent the Duke of Alba with an army to Lisbon to assure his succession. Though the Duke of Alba and the Spanish occupation, however, was little more popular in Lisbon than in Rotterdam, the combined Spanish and Portuguese empires placed into Philip's hands almost the entirety of the explored New World along with a vast trading empire in Africa and Asia. In 1582, when Philip II moved his court back to Madrid from the Atlantic port of Lisbon where he had temporarily settled to pacify his new Portuguese kingdom, the pattern was sealed, in spite of what every observant commentator privately noted: "Sea power is more important to the ruler of Spain than any other prince" wrote a commentator, "for it is only by sea power that a single community can be created out of so many so far apart." A writer on tactics in 1638 observed, "The might most suited to the arms of Spain is that which is placed on the seas, but this matter of state is so well known that I should not discuss it, even if I thought it opportune to do so."[58]

Portugal required an extensive occupation force to keep it under control, and Spain was still reeling from the 1576 bankruptcy. In 1584, William the Silent was assassinated by a half-deranged Catholic, and the death of the popular Dutch resistance leader was hoped to bring an end to the war. It did not. In 1586, Queen Elizabeth I of England, sent support to the Protestant causes in the Netherlands and France, and Sir Francis Drake launched attacks against Spanish merchants in the Caribbean and the Pacific, along with a particularly aggressive attack on the port of Cadiz. In 1588, hoping to put a stop to Elizabeth's intervention, Philip sent the Spanish Armada to attack England. Favourable weather, more heavily armed and manœuverable English ships, and the fact that the English had been warned by their spies in the Netherlands and were ready for the attack resulted in defeat for the Armada. However the failure of the Drake–Norris Expedition to Portugal and the Azores in 1589 marked a turning point in the on-off 1585–1604 Anglo–Spanish War. The Spanish fleets became more effective in transporting greatly increased quantities of silver and gold from the Americas, while English attacks suffered costly failures.

The defense of Cádiz, by Zurbarán

Spain had invested itself in the religious warfare in France after Henry II's death. In 1589, Henry III, the last of the Valois lineage, died at the walls of Paris. His successor, Henry IV of Navarre, the first Bourbon king of France, was a man of great ability, winning key victories against the Catholic League at Arques (1589) and Ivry (1590). Committed to stopping Henry of Navarre from becoming King of France, the Spanish divided their army in the Netherlands and invaded France in 1590.

The Spanish Armada leaving the Bay of Ferrol (1588)

The pacification at the time of Philip III

Faced with wars against England, France and the Netherlands, each led by capable leaders, the bankrupted empire found itself competing against strong adversaries. Continuing piracy against its shipping in the Atlantic and the costly colonial enterprises forced Spain to renegotiate its debts in 1596. The crown attempted to reduce its exposure to the different conflicts, first signing the Treaty of Vervins with France in 1598, recognizing Henry IV (since 1593 a Catholic) as king of France, and restoring many of the stipulations of the previous Peace of Cateau-Cambrésis. The Kingdom of England, suffering from a series of repulses at sea and from an endless guerrilla war by Catholics in Ireland, who were supported by Spain, agreed to the Treaty of London, 1604, following the accession of the more tractable Stuart King James I.

Castile provided the Spanish crown with most of its revenues and its best troops.[59] The plague devastated Castilian lands between 1596 and 1602, causing the deaths of some 600,000 people.[60] A great number of Castilians went to America or died in battle. In 1609, the great majority of the Morisco population of Spain was expelled. It is estimated that Castile lost about 25% of its population between 1600 and 1623. Such a dramatic drop in the population meant the basis for the Crown's revenues was dangerously weakened in a time when it was engaged in continuous conflict in Europe.[61]

Peace with England and France gave Spain an opportunity to focus its energies on restoring its rule to the Dutch provinces. The Dutch, led by Maurice of Nassau, the son of William the Silent and perhaps the greatest strategist of his time, had succeeded in taking a number of border cities since 1590, including the fortress of Breda. Following the peace with England, the new Spanish commander Ambrogio Spinola, a general with the ability to match Maurice, pressed hard against the Dutch and was prevented from conquering the Netherlands only by Spain's latest bankruptcy in 1607. In 1609, the Twelve Years' Truce was signed between Spain and the United Provinces. At last, Spain was at peace — the *Pax Hispanica.*

Spain made a fair recovery during the truce, putting its finances in order and doing much to restore its prestige and stability in the run-up to the last truly great war in which she would play a leading part. Philip II's successor, Philip III, was a man of limited ability, uninterested in politics and preferring to delegate management of the empire to others. His chief minister was the capable Duke of Lerma.

The Duke of Lerma (and to a large extent Philip II) had been uninterested in the affairs of their ally, Austria. In 1618, the king replaced him with Don Balthasar de Zúñiga, a veteran ambassador to Vienna. Don Balthasar believed that the key to restraining the resurgent French and eliminating the Dutch was a closer alliance with Habsburg Austria. In 1618, beginning with the Defenestration of Prague, Austria and the Holy Roman Emperor, Ferdinand II, embarked on a campaign against the Protestant Union and Bohemia. Don Balthasar encouraged Philip to join the Austrian Habsburgs in the war, and Spinola, the rising star of the Spanish army in the Netherlands, was sent at the head of the Army of Flanders to intervene. Thus, Spain entered into the Thirty Years' War.

The Surrender of Breda (1625) to Ambrogio Spinola, by Velazquez. This victory came to symbolize the renewed period of Spanish military vigour in the Thirty Years' War.

The road to Rocroi

In 1621, Philip III was succeeded by the considerably more religious Philip IV. The following year, Don Balthasar was replaced by Gaspar de Guzmán, Count-Duke of Olivares, a reasonably honest and able man who believed that the center of all Spain's woes rested in the Netherlands. After certain initial setbacks, the Bohemians were defeated at White Mountain in 1621, and again at Stadtlohn in 1623. The war with the Netherlands was renewed in 1621 with Spinola taking the fortress of Breda in 1625. The intervention of Christian IV of Denmark in the war worried some (Christian was one of Europe's few monarchs who had no worries over his finances), but the victory of the Imperial general Albert of Wallenstein over the Danes at Dessau Bridge and again at Lutter (both in 1626), eliminated that threat.

There was hope in Madrid that the Netherlands might finally be reincorporated into the Empire, and after the defeat of Denmark the Protestants in Germany seemed crushed. France was once again involved in its own instabilities (the famous Siege of La Rochelle began in 1627), and Spain's eminence seemed clear. The Count-Duke Olivares stridently affirmed, "God is Spanish and fights for our nation these days".[62]

Olivares was a man out of time: he realized that Spain needed to reform, and to reform it needed peace. The destruction of the United Provinces of the Netherlands was added to his list of necessities, because at the root of every anti-Habsburg coalition there was Dutch money. Dutch bankers financed the East India merchants of Seville, and everywhere in the world Dutch entrepreneurship and colonists were undermining Spanish and Portuguese hegemony.

While Spinola and the Spanish army were focused on the Netherlands, the war seemed to go in Spain's favor. But 1627 saw the collapse of the Castilian economy. The Habsburgs had been debasing their currency to pay for the war and prices exploded, just as they had in previous years in Austria. Until 1631, parts of Castile operated on a barter economy owing to the currency crisis, and the government was unable to collect any meaningful taxes from the peasantry and had to depend on revenue from its colonies. The Spanish armies, like others in German territories, resorted to "paying themselves" on the land.

Battle of Nördlingen (1634). Decisive victory for the Catholic Imperial army and Spain over the Swedes.

Olivares had backed certain taxation reforms in Spain pending the end of the war, but was blamed for another embarrassing and fruitless war in Italy. The Dutch, who during the Twelve Years' Truce had made increasing their navy a priority, (which showed its maturing potency at the Battle of Gibraltar 1607), managed to strike a great blow against Spanish maritime trade with the capture of the treasure fleet by captain Piet Hein, on which Spain had become dependent after the economic collapse.

Spanish military resources were stretched across Europe and also at sea as they sought to protect maritime trade against the greatly improved Dutch and French fleets, while still occupied with the Ottoman and associated Barbary pirate threat in the Mediterranean. In the meantime the aim of choking Dutch shipping was carried out by the Dunkirkers with considerable success. In 1625 a Spanish-Portuguese fleet, under Admiral Fradique de Toledo, regained the strategically vital Brazilian city of Salvador da Bahia from the Dutch. Elsewhere, the isolated and undermanned Portuguese forts in Africa and the Asia proved vulnerable to Dutch and English raids and takeovers or simply being bypassed as important trading posts.

In 1630, Gustavus Adolphus of Sweden, one of history's most noted commanders, landed in Germany and relieved the port of Stralsund, the last continental stronghold of German forces belligerent to the Emperor. Gustavus then marched south and won notable victories at Breitenfeld and Lützen, attracting more Protestant support with every step he took. The situation for the Catholics improved with Gustavus's death at Lutzen in 1632, and a key victory at Nordlingen was won in 1634. From a position of strength, the Emperor approached the war-weary German states with a peace in 1635: many accepted, including the two most powerful, Brandenburg and Saxony. Then France entered the equation, and diplomatic calculations were thrown in to confusion.

Cardinal Richelieu of France had been a strong supporter of the Dutch and Protestants since the beginning of the war, sending funds and equipment in an attempt to stem Habsburg strength in Europe. Richelieu decided that the recently signed Peace of Prague was contrary to French designs and declared war on the Holy Roman Emperor and Spain within months of the peace being signed. In the war that followed, the more experienced Spanish forces scored initial successes. Olivares ordered a lightning campaign into northern France from the Spanish Netherlands, hoping to shatter the resolve of King Louis XIII's ministers and topple Richelieu. In the *"année de Corbie"*, 1636, Spanish forces advanced as far south as Corbie, and such was the threat to Paris that the war came close to a conclusion on Spanish terms.

The Battle of Rocroi (1643), the symbolic end of Spain's grandeur; the slow decline sets in.

After 1636, however, Olivares halted the advance, fearful of provoking another crown bankruptcy. The hesitation in pressing home the advantage proved fateful; French forces regrouped and pushed the Spanish back towards the border. The Spanish army would never again penetrate so far. At the Battle of the Downs in 1639 a Spanish fleet carrying troops was destroyed by the Dutch navy, and the Spanish found themselves unable to supply and reinforce their forces adequately in the Netherlands. The Army of Flanders, which represented the finest of Spanish soldiery and leadership, faced a French assault led by Louis II de Bourbon, Prince de Condé in northern France at Rocroi in 1643. The Spanish, led by Francisco de Melo, were beaten by the French. This battle was not a slaughter by any means, however; it was a closely fought battle but the Spanish were forced to surrender with honorable terms. The high reputation of the Army of Flanders was broken at Rocroi, and with it, the grandeur of Spain.

Africa and the Mediterranean

By the 16th century, the Ottomans had become an existential threat to Europe. Ottoman conquests in Europe made significant gains with a decisive victory at Mohács.[63] Charles had preferred to suppress the Ottomans through a considerably more maritime strategy, hampering Ottoman landings on the Venetian territories in the Eastern Mediterranean.

The coastal villages and towns of Spain, Italy and the Mediterranean islands were frequently attacked by Barbary pirates from North Africa; the Formentera was even temporarily left by its population and long stretches of the Spanish and Italian coasts were almost completely abandoned by their inhabitants. The most famous corsair was the Turkish Barbarossa ("Redbeard"). According to Robert C. Davis, between 1 million and 1.25 million Europeans were captured by North African pirates and sold as slaves in North Africa and Ottoman Empire between the 16th and 19th centuries.[64]

The reign of Charles V saw a decline in the presence of Spain in the North of Africa, even if Tunis and its port, La Goleta, were taken in 1535. One after the other, most of the Spanish possessions were lost: Peñón de Vélez de la Gomera (1522), Santa Cruz de Mar Pequeña (1524), Algiers (1529), Tripoli (1551), Bujia (1554), and La Goleta and Tunis (1569).

Only in response to Barbary pirates' raids on the eastern coast of Spain did Charles lead attacks against Tunis (1535) and Algiers (1541).

The Battle of Lepanto (1571), marked the end of the Ottoman naval supremacy in the Mediterranean Sea.

In 1565, the Spanish defeated an Ottoman landing on the strategic island of Malta, defended by the Knights of St. John. Suleiman the Magnificent's death the following year and his succession by his less capable son Selim the Sot emboldened Philip, and he resolved to carry the war to the sultan himself. In 1571, Spanish and Venetian warships, joined by volunteers across Europe, led by Charles's illegitimate son Don John of Austria annihilated the Ottoman fleet at the Battle of Lepanto, in what is perhaps the most decisive battle in modern naval history. The battle ended the threat of Ottoman naval hegemony in the Mediterranean. This mission marked the height of the respectability of Spain and its sovereign abroad as Philip bore the burden of leading the Counter-Reformation.

The Ottomans recovered soon. They reconquered Tunis in 1574, and they helped to restore an ally, Abu Marwan Abd al-Malik I Saadi, in the throne of Morocco, in 1576. The death of the Persian shah, Tahmasp I was an opportunity for the Ottoman sultan to intervene in that country, so, in 1580 was agreed a truce in the Mediterranean with Philip II.[65]

In the first half of the 17th century, Larache and La Mamora, in the Moroccan Atlantic coast, and the island of Alhucemas, in the Mediterranean, were taken, but during the second half of the 17th century, Larache and La Mamora were also lost.

The New World

Explorers and conquistadors

After Columbus, the Spanish colonization of the Americas was led by a series of warrior-explorers called Conquistadors. The Spanish forces, in addition to significant armament and equestrian advantages, exploited the rivalries between competing Indigenous peoples, tribes, and nations, some of which were willing to form alliances with the Spanish in order to defeat their more-powerful enemies, such as the Aztecs or Incas—a tactic that would be extensively used by later European colonial powers. The Spanish conquest was also facilitated by the spread of diseases (e.g. smallpox), common in Europe but never present in the New World and so no immunities, which reduced the indigenous populations in the Americas. This sometimes caused a labour shortage for plantations and public works and so the colonists informally and gradually, at first, initiated the Atlantic slave trade. (*see Population history of American indigenous peoples*)

Emperor Atahualpa is shown surrounded on his palanquin at the Battle of Cajamarca

One of the most accomplished conquistadors was Hernán Cortés, who leading a relatively small Spanish force but with local translators and the crucial support of thousands of native allies, achieved the Spanish conquest of the Aztec Empire in the campaigns of 1519–1521. This territory later became the Viceroyalty of New Spain, present day Mexico. Of equal importance was the Spanish conquest of the Inca Empire by Francisco Pizarro, which would become the Viceroyalty of Peru. After the conquest of Mexico, rumours of golden cities (Quivira and Cíbola in North America and El Dorado in South America) motivated several other expeditions. Many of those returned without having found their goal, or finding it much less valuable than was hoped. Indeed, the New World colonies only began to yield a substantial part of the Crown's revenues with the establishment of mines such as that of Potosí (Bolivia) and Zacatecas (Mexico) both started in 1546. By the late 16th century, silver from the Americas accounted for $^1/_5$ of Spain's total budget.[66] In the 16th century "perhaps 240,000 Europeans" entered American ports.[67]

Further Spanish settlements were progressively established in the New World: New Granada in the 1530s (later in the Viceroyalty of New Granada in 1717 and present day Colombia), Lima in 1535 as the capital of the Viceroyalty of Peru, Buenos Aires in 1536 (later in the Viceroyalty of the Río de la Plata in 1776), and Santiago in 1541.

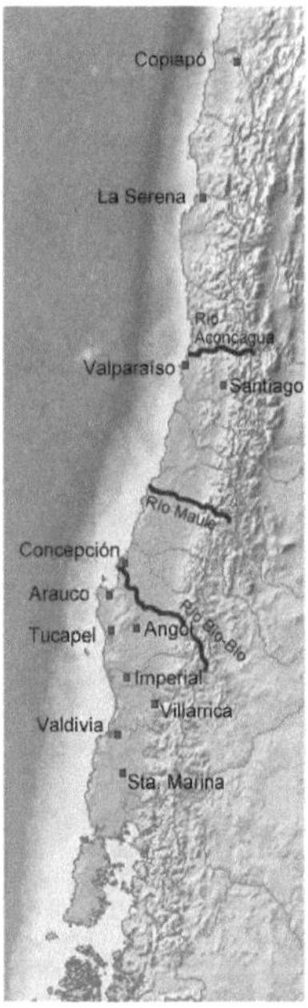

Spanish settlements in Chile before the Destruction of the Seven Cities; in 1604, all settlements south Biobío River except those in Chiloé had fallen.

Florida was colonized in 1565 by Pedro Menéndez de Avilés when he founded Saint Augustine and then promptly defeated an attempt led by the French Captain Jean Ribault and 150 of his countrymen to establish a French foothold in Spanish Florida territory. Saint Augustine quickly became a strategic defensive base for the Spanish ships full of gold and silver being sent to Spain from its New World dominions.

The Portuguese Ferdinand Magellan died while in the Philippines commanding a Castilian expedition to circumnavigate the globe in 1522. Juan Sebastián Elcano would lead the expedition to success. Therefore, Spain sought to enforce their rights in the Moluccan islands, which led a conflict with the Portuguese, but the issue was resolved with the Treaty of Zaragoza (1525), settling the location of the antimeridian of Tordesillas, which would divide the world into two equal hemispheres. Thenceforth, maritime expeditions led to the discovery of several archipelagos in the South Pacific as the Pitcairn Islands, the Marquesas, Tuvalu, Vanuatu, the Solomon Islands or New Guinea.

On April 27, 1565, the first permanent Spanish settlement in the Philippines was founded by Miguel López de Legazpi and the service of Manila Galleons was inaugurated. The Manilla Galleons shipped goods from all over Asia across the Pacific to Acapulco on the coast of Mexico. From there, the goods were transshipped across Mexico to the Spanish treasure fleets, for shipment to Spain. The Spanish trading post of Manila was established to facilitate this trade in 1572. The control of Guam, Mariana Islands, Caroline Islands, and Palau was later, from the end of the 17th century, and remained under Spanish control until 1898.

In 1599, the Spanish Empire suffered one of its greatest setbacks in the Americas when native Mapuches destroyed the Spanish army of Chile and killed the governor Martín García Óñez de Loyola in the Battle of Curalaba. This event led to a generalized rebellion that ended in the Destruction of the Seven Cities, and established the Bío-Bío River as frontier. This led to the prolonged war called the War of Arauco, and southern Chile received the name of *Flandés Indiano* (Indian Flanders) due to the resistances of the Mapuches.[68] In a case unique for America, the Spanish established a professional army in Chile financed through the Real Situado by the Viceroyalty of Peru.

Organization and administration

The Laws of the Indies resulted in the Laws of Burgos, 1512–1513, which were the first codified set of laws governing the behavior of Spanish settlers in the Americas, particularly with regards to treatment of native Indians. They forbade the maltreatment of natives, and endorsed the Indian Reductions with attempts of conversion to Catholicism.[69] Upon their failure, they were replaced by the New Laws (1542)

Spain passed some laws for the protection of the indigenous peoples of its American colonies, the first such in 1542; the legal thought behind them was the basis of modern international law. Taking advantage of their extreme remoteness, the European colonists revolted when they saw their power being reduced, forcing a partial revoking of these New Laws. Later, weaker laws were introduced to protect the indigenous peoples but records show their effect was limited. The restored *Encomenderos* increasingly used native Indian workforce.

The Last Spanish Habsburgs (1643–1700)

Traditionally, historians mark the Battle of Rocroi (1643) as the end of Spanish dominance in Europe, but the war was not finished. Supported by the French, the Catalans, Neapolitans, and Portuguese rose up in revolt against the Spanish in the 1640s. With the Spanish Netherlands caught between the tightening grip of French and Dutch forces after the Battle of Lens in 1648, the Spanish made peace with the Dutch and recognized the independent United Provinces in the Peace of Westphalia that ended both the Eighty Years' War and the Thirty Years' War.

War with France continued for eleven more years. Although France suffered from a civil war from 1648–1652 (*see Wars of the Fronde*) the Spanish economy was so exhausted that it was unable to effectively cope with war on so many fronts. Yet the decline of Spanish power in this period has often been overstated. Spain retook Naples in 1648 and Catalonia in 1652, but the war came to an end at the Battle of the Dunes (1658) where the French army under Viscount Turenne defeated the remnants of the Spanish army of the Netherlands. Spain agreed to the Peace of the Pyrenees in 1659 that ceded to France Roussillon and Artois.

Meeting of Philip IV of Spain and Louis XIV of France on 7 July 1660 at Pheasant Island.

Portugal had rebelled in 1640 under the leadership of John of Braganza, a pretender to the throne. He had received widespread support from the Portuguese people, and Spain—which had to deal with rebellions elsewhere, along with the war against France – was unable to respond adequately. John mounted the throne as King John IV of Portugal and the Spanish and Portuguese co-existed in a de facto state of peace from 1644–1656. When John died in 1656, the Spanish attempted to wrest Portugal from his son Alfonso VI of Portugal but were defeated at Ameixial (1663) and Montes Claros (1665), leading to Spain's recognition of Portugal's independence in 1668.

Spain still had a huge overseas empire, but France was now the superpower in Europe and the United Provinces were in the Atlantic.

The Great Plague of Seville (1647–1652) killed up to 25% of Seville's population. Sevilla, and indeed the economy of Andalucía, would never recover from so complete a devastation. Altogether Spain was thought to have lost 500,000 people, out of a population of slightly fewer than 10,000,000, or nearly 5% of its entire population. Historians reckon the total cost in human lives due to these plagues throughout Spain, throughout the entire 17th century, to be a minimum of nearly 1.25 million.[70]

The regency of the young Spanish king Charles II was incompetent in dealing with the War of Devolution that Louis XIV of France prosecuted against the Spanish Netherlands in 1667–68, losing considerable prestige and territory, including the cities of Lille and Charleroi. In the Franco-Dutch War of 1672–1678, Spain lost still more territory when it came to the assistance of its former Dutch enemies, most notably Franche-Comté. In the Nine Years' War (1688–1697) Louis once again invaded the Spanish Netherlands. French forces led by the Duke of Luxembourg defeated the Spanish at Fleurus (1690), and subsequently defeated Dutch forces under William III of Orange, who fought on Spain's side. The war ended with most of the Spanish Netherlands under French occupation, including the important cities of Ghent and Luxembourg. The war revealed to Europe how vulnerable and backward the Spanish defenses and bureaucracy were, but the ineffective Spanish Habsburg government took no action to improve them.

The final decades of the 17th century saw utter decay and stagnation in Spain; while the rest of Western Europe went through exciting changes in government and society — the Glorious Revolution in England and the reign of the Sun King in France — Spain remained adrift. The Spanish bureaucracy that had built up around the charismatic, industrious, and intelligent Charles I and Philip II demanded a strong and hardworking monarch; the weakness and lack of interest of Philip III and Philip IV contributed to Spain's decay. Charles II was mentally retarded and impotent. He was therefore childless, and in his final will he left his throne to the Bourbon prince Philip of Anjou, rather than to a fellow Habsburg, albeit from Austria. This resulted in the War of the Spanish Succession.

The Bourbon Spanish Empire: reform and recovery (1700–1808)

Under the Treaties of Utrecht (April 11, 1713), the European powers decided what the fate of Spain would be, in terms of the continental balance of power. The new Bourbon king Philip V retained the Spanish overseas empire, but ceded the Spanish Netherlands, Naples, Milan, and Sardinia to Austria; Sicily and parts of Milan to Duchy of Savoy; and Gibraltar and Minorca to the Kingdom of Great Britain. Moreover, Philip V granted the British the exclusive right to slave trading in Spanish America for thirty years, the so-called *asiento*, as well as licensed voyages to ports in Spanish colonial dominions, openings, as Fernand Braudel remarked, for both licit and illicit smuggling (Brudel 1984 p 418).

Spain's economic and demographic recovery had begun slowly in the last decades of the Habsburg reign, as was evident from the growth of its trading convoys and much more rapid growth of illicit trade during the period, though this growth was slower than in its northern rivals who had gained increasing illicit access to its empire's markets. Critically, this recovery was not translated into institutional improvement because of the incompetent leadership of the unfortunate last Habsburg. This legacy of neglect was reflected in the early years of Bourbon rule in which the military was ill-advisedly pitched into battle against the Quadruple Alliance (1718–1720).

The *Battle of Cape Passaro*, 11 August 1718.

Following the war, the new Bourbon monarchy would take a much more cautious approach to international relations, built upon a family alliance with Bourbon France, and continuing to follow a program of institutional renewal.

Bourbon Reforms

At the beginning of the Philip V's reign and due to the War of the Spanish Succession, the Spanish king Philip V initiated organizational reforms headed for a government more executive, giving priority to the direct decision of the monarch, opposite to the deliberative way of the polisynodial system of Councils.[71]

The Spanish Bourbons' broadest intentions were to break the power of the entrenched aristocracy of the *Criollos* in America (locally born colonials of European descent), and, eventually, loosen the territorial control of the Society of Jesus over the virtually independent theocracies of Guarani *Misiones*: the Jesuits were expelled from Spanish America in 1767. In addition to the established *consulados* of Mexico City and Lima, firmly in the control of local landowners, a new rival *consulado* was set up at Vera Cruz.

Immediately, Philip's government set up a ministry of the Navy and the Indies (1714) and created first a Honduras Company (1714), a Caracas company, the Guipuzcoana Company (1728) and—the most successful one—a Havana Company (1740). In 1717–1718, the structures for governing the Indies, the *Consejo de Indias* and the *Casa de Contratación* that governed investments in the cumbersome escorted fleets were transferred from Seville to Cádiz, which became the one port for all Indies trading (see flota system). Individual sailings at regular intervals were slow to displace the old habit of armed convoys, but by the 1760s there were regular packet ships plying the Atlantic between Cádiz and Havana and Puerto Rico, and at longer intervals to the Río de la Plata, where an additional viceroyalty was created in 1776. The contraband trade that was the lifeblood of the Habsburg empire declined in proportion to registered shipping (a shipping registry having been established in 1735).

Two upheavals registered unease within Spanish America and at the same time demonstrated the renewed resiliency of the reformed system: the Tupac Amaru uprising in Peru in 1780 and the rebellion of the *comuneros* of New Granada, both in part reactions to tighter, more efficient control.

18th century prosperity

The 18th century was a century of prosperity for the overseas Spanish Empire as trade within grew steadily, particularly in the second half of the century, under the Bourbon reforms. Spain's crucial victory in the Battle of Cartagena de Indias against an extraordinary British fleet, in the Caribbean port of Cartagena de Indias, one of a number of successful battles, helped it secure Spain's dominance of the Americas until the 19th century.

San Felipe de Barajas Fortress Cartagena de Indias. In 1741 the Spanish defeated a vast British invasion fleet and army from this fortress in present-day Colombia during the Battle of Cartagena de Indias.

With a Bourbon monarchy came a repertory of Bourbon mercantilist ideas based on a centralized state, put into effect in America slowly at first but with increasing momentum during the century. Rapid shipping growth from the mid-1740s until the Seven Years' War (1756–1763), reflecting in part the success of the Bourbons in bringing illicit trade under control. With the loosening of trade controls after the Seven Years War, shipping trade within the empire once again began to expand, reaching an extraordinary rate of growth in the 1780s.

The ending of Cádiz's trade monopoly with America brought about a rebirth of Spanish manufactures. Most notable was the rapidly growing textile industry of Catalonia which by the mid-1780s saw the first signs of industrialisation. This saw the emergence of a small, politically active commercial class in Barcelona. This isolated pocket of advanced economic development stood in stark contrast to the relative backwardness of most of the country. Most of the improvements were in and around some major coastal cities and the major islands such as Cuba, with its plantations, and a renewed growth of precious metals mining in America. On the other hand most of rural Spain and its empire, where the great bulk of the population lived, lived in relatively backward conditions by 18th century West European standards, reinforced old customs and isolation. Agricultural productivity remained low despite efforts to introduce new techniques to what was for the most part an uninterested, exploited peasant and labouring groups. Governments were inconsistent in their policies. Though there were substantial improvements by the late 18th century, Spain was still an economic backwater. Under the mercantile trading arrangements it had difficulty in providing the goods being demanded by the strongly growing markets of its empire, and providing adequate outlets for the return trade.

Overseas expansion

A Spanish army captures British Pensacola in 1781. In 1783 the Treaty of Paris returns all of Florida to Spain for the return of the Bahamas.

The Bourbon institutional reforms were to bear some fruit militarily when Spanish forces easily retook Naples and Sicily from the Austrians in 1734 (War of the Polish Succession) and thwarted British campaigns attempting to seize the strategic cities of Cartagena de Indias and Cuba during the War of Jenkins' Ear (1739–1742). Moreover, though Spain lost territories to greatly improved and successful amphibious British forces towards the end of the Seven Years' War (1756–1763), she was to recover these losses and seize the British naval base in the Bahamas during the American Revolutionary War (1775–83).

The greater part of what is the territory of today's Brazil had been claimed as Spanish when exploration began with the navigation of the length of the Amazon River in 1541–42 by Francisco de Orellana. Many Spanish expeditions explored large parts of this vast region, especially those areas close to the main areas of Spanish settlement. During the 16th and 17th centuries, Spanish soldiers, missionaries and adventurers also established pioneering communities, primarily in Paraná, Santa Catarina, São Paulo and forts on the northeastern coast threatened by the French and Dutch. As Portuguese-Brazilian settlement expanded, following in the trail of the Bandeirantes exploits, these isolated Spanish groups were eventually integrated into Brazilian society. Only some Castilians who were displaced from the disputed areas of the Pampas of Rio Grande do Sul have left a significant influence on the formation of the gaucho, when they mixed with Indian groups, Portuguese and blacks who arrived in the region during the 18th century. The Spanish were barred by their laws from slaving of indigenous people, leaving them without a commercial interest deep in the interior of the Amazon basin. The Laws of Burgos (1512) and the New Laws (1542) had been intended to protect the interests of indigenous people. While in spirit they were often abused, as through forced exploitative labour of locals, they did prevent widespread formal enslavement of indigenous people in Spanish territories. The Protuguese-Brazilian slavers, the Bandeirantes, had the advantage of access from the mouth of the Amazon River, which was on the Portuguese side of the line of Tordesillas. One famous attack upon a Spanish mission in 1628 resulted in the enslavement of about 60,000 indigenous people.[72] In time, there were in effect a self funding force of occupation. By the 18th century, much of the Spanish territory was under defacto control of Portuguese-Brazil. This reality was recognised with the legal transfer of sovereignty in 1750 of most of the Amazon basin and surrounding areas to Portugal in the Treaty of Madrid. This settlement sowed the seeds of the Guarani War in 1756.

The California mission planning was begun in 1769. The Nootka Crisis (1789–1791) involved a dispute between Spain and Great Britain about the British settlement in Oregon to British Columbia. In 1791, the king of Spain gave Alessandro Malaspina an order to search for a Northwest Passage.

The Spanish empire had still not returned to first rate power status, but it had recovered considerably from the dark days at the beginning of the 18th century when it was, and particularly in continental matters, at the mercy of other powers' political deals. The relatively more peaceful century under the new monarchy had allowed it to rebuild and start the long process of modernizing its institutions and economy. The demographic decline of the 17th century had been reversed. It was a middle ranking power with great power pretensions that could not be ignored. But time was to be against it. The growth of trade and wealth in the colonies caused increasing political tensions as frustration grew with the improving but still restrictive trade with Spain. Malaspina's recommendation to turn the empire into a looser confederation to help improve governance and trade so as to quell the growing political tensions between the élites of the empire's periphery and centre was suppressed by a monarchy afraid of losing control. All was to be swept away by the tumult that was to overtake Europe at the turn of the 19th century with the French Revolutionary and Napoleonic Wars.

Twilight Of The Global Empire (1800–1899)

The first major territory Spain was to lose in the 19th century was the vast and wild Louisiana Territory, which stretched north to Canada and was ceded by France in 1763 under the terms of the Treaty of Fontainebleau. The French, under Napoleon, took back possession as part of the Treaty of San Ildefonso in 1800 and sold it to the United States (Louisiana Purchase, 1803).

Churruca's Death, oil on canvas about the Battle of Trafalgar by Eugenio Álvarez Dumont, Prado Museum

The destruction of the main Spanish fleet, under French command, at the Battle of Trafalgar (1805) undermined Spain's ability to defend and hold on to its empire. The later intrusion of Napoleonic forces into Spain in 1808 (see Peninsular War) cut off effective connection with the empire. But it was internal tensions that ultimately ended the empire in the Americas.

Napoleon's sale in 1803 of the Louisiana Territory to the United States caused border disputes between the United States and Spain that, with rebellions in West Florida (1810) and in the remainder of Louisiana at the mouth of the Mississippi, led to their eventual cession to the United States, along with the sale of all of Florida, in the Adams–Onís Treaty (1819). In 1806 Baron Nikolai Rezanov attempted to negotiate a treaty between the Russian-American Company and Viceroyalty of New Spain but his premature death in 1807 ended any treaty hopes.

The Second of May, 1808: The Charge of the Mamelukes, by Francisco de Goya (1814).

In 1808, Napoleon forced the abdication of the Spanish monarchy and placed his brother on the throne, however his unpopularity provoked an uprising from the Spanish people and the grinding guerrilla warfare, which Napoleon dubbed his "ulcer", the Peninsular War, (famously depicted by the painter Goya) ensued, followed by a power vacuum lasting up to a decade and turmoil for several decades, civil wars on succession disputes, a republic, and finally a liberal democracy. Spain lost all the colonial possessions in the first third of the century, except for Cuba, Puerto Rico and, isolated on the far side of the globe, the Philippines, Guam and nearby Pacific islands, as well as Ceuta and Melilla, and some islands that later would be part of the Spanish Guinea.

The wars of independence in Spanish America were triggered by a British attempt to seize the Viceroyalty of the Río de la Plata in 1806. The viceroy retreated hastily to the hills when defeated by a small British force. However when the *Criollos* militias and colonial army defeated the now reinforced British force in 1807, and with the example of the North American revolutionaries very much in their minds, they quickly set about the business of winning their own independence and inspiring independence movements elsewhere in America. A long period of wars began which led to the independence of Paraguay (1811) and Uruguay (1815 but subsequently ruled by Brazil until 1828). José de San Martín campaigned for freedom in Argentina (1816), Chile (1818) and Peru (1821). Further north Simón Bolívar led forces that won independence for the area that is currently Venezuela, Colombia (then Nueva Granada), Ecuador, and Bolivia (then Alto Perú) by 1825. In 1810, a free thinking priest, Miguel Hidalgo y Costilla, declared Mexican independence, which was won by 1821. Central America declared its independence in 1821 and was joined to Mexico for a brief time (1822–23). Panama declared independence in 1821 and joined Nueva Granada, Ecuador and Venezuela to form the Republic of Gran Colombia (from 1821–1903). Santo Domingo likewise declared independence in 1821 and began negotiating for inclusion in Bolivar's Republic of Gran Colombia, but was quickly occupied by Haiti, which ruled it until an 1844 revolution. Then after 17 years of independence, in 1861, Santo Domingo was again made a colony due to Haitian aggressions, yet by 1865 Santo Domingo again declared independence, making it the only territory to be recolonized by Spain. Thus only Cuba and Puerto Rico remained in Spanish hands in the New World after 1865.

The Battle of Ayacucho, Peru May 1824

In devastated Spain, the post-Napoleonic era created a political vacuum, broke apart any traditional consensus on sovereignty, fragmented the country politically and regionally and unleashed wars and disputes between progressives, liberals and conservatives. The instability inhibited Spain's development, which had started fitfully gathering pace in the previous century. A brief period of improvement occurred in the 1870s when the capable Alfonso XII of Spain and his thoughtful ministers succeeded in restoring some vigour to Spanish politics and prestige, but this was cut short by Alfonso's early death.

Vizcaya explodes in the Battle of Santiago de Cuba

An increasing level of nationalist, anti-colonial uprisings in various colonies culminated with the Spanish–American War of 1898, fought primarily over Cuba. Military defeat was followed by the independence of Cuba and the cession, for US$20 million, of Puerto Rico, the Philippines, and Guam to the United States. On June 2, 1899, the last Spanish garrison in the Philippines, located in Baler, Aurora, was pulled out, effectively ending around 300 years of Spanish hegemony in this archipelago. Its American and Asian presence ended, Spain then sold its remaining Pacific Ocean possessions to Germany in 1899, retaining only its African territories.

Territories in Africa (1885–1975)

By the end of the 17th century, only Melilla, Alhucemas, Peñón de Vélez de la Gomera (which had been taken again in 1564), Ceuta (part of the Portuguese Empire since 1415, has chosen to retain its links to Spain once the Iberian Union ended; the formal allegiance of Ceuta to Spain was recognized by the Treaty of Lisbon in 1668), Oran and Mazalquivir remained as Spanish territory in Africa. The latter cities were lost in 1708, reconquered in 1732 and sold by Charles IV in 1792.

In 1778, Fernando Poo Island (now Bioko), adjacent islets, and commercial rights to the mainland between the Niger and Ogooué Rivers were ceded to Spain by the Portuguese in exchange for territory in South America (Treaty of El Pardo). In the 19th century, some Spanish explorers and missionaries would cross this zone, among them Manuel de Iradier.

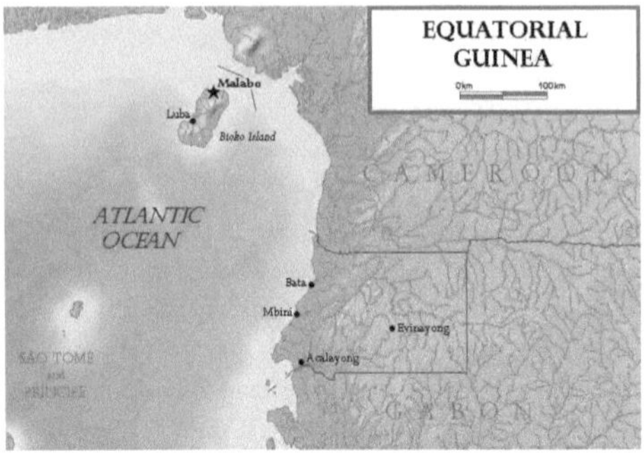

Map of Equatorial Guinea

In 1848, Spanish troops conquered the Islas Chafarinas.

In 1860, after the Tetuan War, Morocco ceded Sidi Ifni to Spain as a part of the Treaty of Tangiers, on the basis of the old outpost of Santa Cruz de la Mar Pequeña, thought to be Sidi Ifni. The following decades of Franco-Spanish collaboration resulted in the establishment and extension of Spanish protectorates south of the city, and Spanish influence obtained international recognition in the Berlin Conference of 1884: Spain administered Sidi Ifni and Western Sahara jointly. Spain claimed a protectorate over the coast of Guinea from Cape Bojador to Cap Blanc, too. Río Muni became a protectorate in 1885 and a colony in 1900. Conflicting claims to the Guinea mainland were settled in 1900 by the Treaty of Paris.

Following a brief war in 1893, Spain expanded its influence south from Melilla.

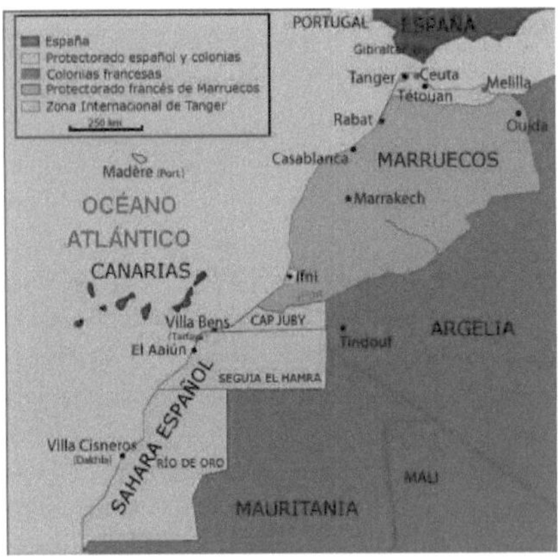

Morocco

In 1911, Morocco was divided between the French and Spanish. The Rif Berbers rebelled, led by Abdelkrim, a former officer for the Spanish administration. The *Battle of Annual* (1921) was a sudden, grave, and almost fatal, military defeat suffered by the Spanish army against Moroccan insurgents. A leading Spanish politician emphatically declared: "*We are at the most acute period of Spanish decadence*". The statement reflected the mood of the country. The rebellion exposed the utter corruption and incompetence of the military and destabilised the Spanish government, leading to dictatorship. A campaign in conjunction with the French suppressed the Rif rebels by 1925 but at a terrible cost to both sides. In 1923, Tangier was declared an international city under French, Spanish, British, and later Italian joint administration. The African army, led by a veteran of the Moroccan campaign, Francisco Franco, started the Spanish Civil War (1936–39). Between 1926 and 1959, Bioko and Rio Muni were united as the colony of Spanish Guinea. During the Second World War the Vichy French presence in Tangier was overcome by that of Francoist Spain.

Spain lacked the wealth and the interest to develop an extensive economic infrastructure in its African colonies during the first half of the 20th century. However, through a paternalistic system, particularly on Bioko Island, Spain developed large cocoa plantations for which thousands of Nigerian workers were imported as laborers. The Spanish also helped Equatorial Guinea achieve one of the continent's highest literacy rates and developed a good network of health care facilities.

In 1956, when French Morocco became independent, Spain surrendered Spanish Morocco to the new nation, but retained control of Sidi Ifni, the Tarfaya region and Spanish Sahara. Moroccan Sultan (later King) Mohammed V was interested in these territories and invaded Spanish Sahara in 1957 (The Ifni War, or, in Spain, the Forgotten War, *la Guerra Olvidada*). In 1958, Spain ceded Tarfaya to Mohammed V and joined the previously separate districts of Saguia el-Hamra (in the north) and Río de Oro (in the south) to form the province of Spanish Sahara.

In 1959, the Spanish territory on the Gulf of Guinea was established with a status similar to the provinces of metropolitan Spain. As the Spanish Equatorial Region, it was ruled by a governor general exercising military and civilian powers. The first local elections were held in 1959, and the first Equatoguinean representatives were seated in the Spanish parliament. Under the Basic Law of December 1963, limited autonomy was authorized under a joint legislative body for the territory's two provinces. The name of the country was changed to Equatorial Guinea.

In March 1968, under pressure from Equatoguinean nationalists and the United Nations, Spain announced that it would grant the country independence. At independence, Equatorial Guinea had one of the highest per capita incomes in Africa. In 1969, under international pressure, Spain returned Sidi Ifni to Morocco. Spanish control of Spanish Sahara endured until the 1975 Green March prompted a withdrawal, under Moroccan military presion. The future of this former Spanish colony remains uncertain.

The Canary Islands and Spanish cities in the African mainland are considered an equal part of Spain and the European Union but have a different tax system without Value Added Tax.

Morocco still claims Ceuta, Melilla, and *plazas de soberanía* even though they are internationally recognized as administrative divisions of Spain (despite Plazas de Soberania which is a territory of Spain). Isla Perejil (Arabic: Leila *("night")) was occupied on July 11, 2002 by Moroccan Gendarmerie and troops, who were evicted peacefully by Spanish naval forces.*

Legacy

The Spanish language (now the second most widely spoken language in the world) and the Roman Catholic faith were brought to the Americas, parts of Africa and the Spanish East Indies, by Spanish colonization which began in the 15th century. It also played a crucial part in sustaining the Catholic Church as the leading Christian denomination in Europe when it was under extreme pressure.

The Renaissance Cathedral of Lima is a legacy of the Spanish settlement in that city.

The long colonial period in Spanish America resulted in a mixing of peoples. Most Hispanics in the Americas have mixed American Indian and European ancestry, while a substantial proportion also have African ancestry. The only exceptions are Argentina and Uruguay which experienced heavy European immigration in the post colonial period.

In concert with the Portuguese Empire, the Spanish Empire laid the foundations of a truly global trade by opening up the great trans-oceanic trade routes. The Spanish dollar became the world's first global currency.

One of the features of this trade was the exchange of a great array of domesticated plants and animals between the Old World and the New and vice versa. Some that were introduced to the Americas included wheat, barley, apples, cattle, sheep, pigs, horses, donkeys, among many others. The Old World received from the Americas such things as maize, potatoes, chili peppers, tomatoes and tobacco, to name just a few examples. The result of these exchanges was to significantly improve the agricultural potential of not only the Americas, but also that of Europe and Asia.

There were also cultural influences, which can be seen in everything from architecture to food, music, art and law, from Argentina to the Southwest United States. The complex origins and contacts of different peoples resulted in cultural influences coming together in the varied forms so evident today in the former colonial areas.

See also

- History of Spain
- List of Spanish wars
- Panhispanism
- Spanish Guinea
- Imperator totius Hispaniae
- List of largest empires

References

Notes

[1] Raudzens, George (1999), *Empires: Europe and globalization, 1492-1788*, Sutton, p.68 (http://books.google.es/books?ei=CLYoTsi1F5TB8QPfj5ytAw&ct=result&id=8K9mAAAAMAAJ&dq=Spain+reached+the+peak+of+its+world+power&q=1556#search_anchor)

[2] The Mongol Empire had been larger, but was restricted to Eurasia

[3] Farazmand, Ali (1994). *Handbook of bureaucracy*. M. Dekker. pp. 12–13. ISBN 9780824791827.

[4] Ruiz Martín, Felipe (1996). *La proyección europea de la monarquía hispánica* (http://books.google.es/books?id=jlJPsVc8UvQC&lpg=PA473&lr&pg=PA473#v=onepage&q&f=false). Editorial Complutense. p. 473. ISBN 9788495983305. .

[5] Denominations as Iberian union (http://books.google.com/books?id=hhNfVshMw64C&pg=PA495&dq="Iberian+union"&lr=&as_brr=3&hl=es), imperio hispano-portugués (http://books.google.com/books?id=-REq5SekR2kC&pg=PA42&dq=imperio+hispano-portuguÃ©s&lr=&hl=es), Spanish-Portuguese empire (http://books.google.com/books?id=aegXU82-HO4C&pg=PA21&dq="Spanish-Portuguese+Empire"&as_brr=3&hl=es), dual monarchy (http://books.google.com/books?id=8RpIxe2utj8C&pg=PA42&dq=monarchy+dual+1580&lr=&as_brr=3&hl=es), Portugal as part of the Spanish Monarchy (http://books.google.es/books?id=X2Hf-crzPjUC&pg=PA543&vq=monarchy+spanish+portugal&dq=kingdoms+of+the+monarchy+spanish&lr=&as_brr=3&as_pt=ALLTYPES&source=gbs_search_s&cad=0), Portugal incorporated into the Habsburg monarchy (http://books.google.es/books?id=59W7fW6IraIC&pg=PA36&dq=portugal+monarchy+1580&lr=#PPA36,M1), Portugal incorporated in the Spanish Monarchy (http://books.google.es/books?id=aRlBTFGz3SUC&pg=PA36&dq=portugal+monarchy+1580&lr=), Habsburg rule in Portugal (http://books.google.es/books?id=C0lrDYYzMyAC&pg=PR14&dq=habsburg+rule+portugal&lr=&as_brr=3&as_pt=ALLTYPES), or union of Castile and Portugal (http://books.google.es/books?id=gbU8AAAAIAAJ&pg=PA462&dq=union++portugal+castile+1580&lr=&as_brr=3&as_pt=ALLTYPES)

[6] John Huxtable Elliott (2002) *España en Europa: Estudios de historia comparada: escritos seleccionados*, page 80 (http://books.google.es/books?id=h-rlHXgCnAIC&pg=PA80&vq=uniÃ³n+dinÃ¡stica+aeque+principaliter&dq=espaÃ±a+portugal+castilla&source=gbs_search_s&cad=0)

[7] Jean-Frédéric Schaub (2001) *Le Portugal au temps du Comte-Duc d'Olivares, 1621–1640*, pag 59 (http://books.google.es/books?id=ylbW1HcUTfsC&pg=PA59&vq=portugal+union++monarchie&dq=kingdoms+of+the+monarchy+spanish&lr=&as_brr=3&as_pt=ALLTYPES&source=gbs_search_s&cad=0)

[8] Ali Farazmand (1994) *Handbook of Bureaucracy*, page 13 (http://books.google.es/books?id=8NBc_QT26ZoC&pg=PA13&vq=castile+aragon+catalonia+valencia+portugal+sicily+naples+milan&dq=spanish+empire&lr=&as_brr=3&source=gbs_search_s&cad=0)

[9] Wolfgang Reinhard, European Science Foundation (1996), *Power Elites and State Building*, pag 92 (http://books.google.es/books?id=et6x39Ws-j4C&pg=PA92&vq=spanish+monarchy+Castile+Aragon+Portugal&dq=kingdoms+of+the+monarchy+spanish&lr=&as_brr=3&as_pt=ALLTYPES&source=gbs_search_s&cad=0)

[10] Daviken Studnicki-Gizbert (2007), *A Nation Upon the Ocean Sea: Portugal's Atlantic Diaspora and the Crisis of the Spanish Empire, 1492–1640*, page 36 (http://books.google.es/books?id=59W7fW6IraIC&pg=PA36&vq=Portugal+Spanish+empire&dq=portugal+monarchy+1580&lr=&source=gbs_search_s&cad=0)

[11] Anderson 2000, p. 103

[12] Lockhart & Schwartz 1983, p. 250

[13] Lach & Van Kley 1994, p. 9

[14] Donald F. Lach, Edwin J. Van Kley (1993), *Asia in the Making of Europe: A Century of Advance*, page 9 (http://books.google.com/books?id=UJJVYZ0q8UAC&pg=PA9&vq=iberian+union+Spanish+portuguese+empire&dq=iberian+union+empire&hl=es&source=gbs_search_s&cad=0)

[15] Kamen 2003, p. 403

[16] Castañeda Delgado, Paulino (1996), "La Santa Sede ante las empresas marítimas ibéricas" (http://www.bibliojuridica.org/libros/2/725/14.pdf), *La Teocracia Pontifical en las controversias sobre el Nuevo Mundo*, Universidad Autónoma de México, ISBN 968-36-5153-4, .

[17] Newitt, Malyn (2005), *A history of Portuguese overseas expansion, 1400–1668* (http://books.google.es/books?id=eyQu-O1EQCAC&lpg=PA38&dq=&pg=PA38#v=onepage&q=&f=false), Routledge, pp. 38, ISBN 99780415239806, .

[18] Bakewell, Peter (2009), *A History of Latin America to 1825* (http://books.google.es/books?id=MTlK0Mx1a38C&lpg=PA82&dq=&pg=PA82#v=onepage&q&f=true), John Wiley and Sons, pp. 82, ISBN 9781405183680, .

[19] Rialp, Ediciones, S.A. (1992), *Historia general de España y América* (http://books.google.es/books?id=4DWBNjs8iwEC&lpg=PA80&dq=&pg=PA80#v=onepage&q&f=false), **10**, Ediciones Rialp, pp. 82, ISBN 9788432121029, .

[20] Soyer, François (2007), *The persecution of the Jews and Muslims of Portugal: King Manuel I and the end of religious tolerance (1496-7)* (http://books.google.es/books?id=hvOpWcWD2_wC&lpg=PA274&dq=&pg=PA274#v=onepage&q&f=false), BRILL, pp. 274, ISBN 9789004162624, .

[21] Davenport, Frances Gardiner (2004), *European Treaties Bearing on the History of the United States and Its Dependencies* (http://books.google.es/books?id=mDPF4ILESaUC&lpg=PA49&dq=&pg=PA49#v=onepage&q=&f=false), The Lawbook Exchange, Ltd., pp. 49, ISBN 9781584774228, .

[22] Burbank, Jane; Frederick Cooper (2010). *Empires in World History: Power and the Politics of Difference* (http://books.google.es/books?id=y7B9euuLEkUC&lpg=PA120&dq=&pg=PA120#v=onepage&q=&f=true). Princeton University Press. pp. 120–121. ISBN 9780691127088. .

[23] Fernández Herrero, Beatriz (1992) (in Spanish). *La utopía de América: teoría, leyes, experimentos* (http://books.google.es/books?id=Lcv7a0tM7XgC&lpg=PA143&dq=&pg=PA143#v=onepage&q=&f=false). Anthropos Editorial. p. 143. ISBN 9788476583203. .

[24] McAlister, Lyle N. (1984). *Spain and Portugal in the New World, 1492-1700* (http://books.google.es/books?id=lFwnCYl85VEC&lpg=PA69&dq=&pg=PA69#v=onepage&q=&f=false). U of Minnesota Press. p. 69. ISBN 9780816612185. .

[25] (in Spanish) *Historia general de España y América* (http://books.google.es/books?id=4DWBNjs8iwEC&lpg=PA190&dq=&pg=PA189#v=onepage&q&f=true). **10**. Ediciones Rialp. 1992. p. 189. ISBN 9788432121029. .

[26] Fernández Herrero, Beatriz (1992) (in Spanish). *La utopía de América: teoría, leyes, experimentos* (http://books.google.es/books?id=Lcv7a0tM7XgC&lpg=PA141&dq=&pg=PA141#v=onepage&q=&f=true). Anthropos Editorial. p. 141. ISBN 9788476583203. .

[27] Vieira Posada, Édgar (2008), *La formación de espacios regionales en la integración de América Latina* (http://books.google.es/books?id=-PAQOJyi-1UC&lpg=PA56&dq=&pg=PA56#v=onepage&q=&f=false), Pontificia Universidad Javeriana, pp. 56, ISBN 9789586982344, .

[28] Sánchez Doncel, Gregorio (1991), *Presencia de España en Orán (1509–1792)* (http://books.google.es/books?id=EOV8qNnYvDwC&lpg=PA129&dq=&pg=PA122#v=onepage&q&f=true), I.T. San Ildefonso, pp. 122, ISBN 9788460076148, .

[29] Rialp, Ediciones, S.A. (1981), *Los Trastámara y la Unidad Española* (http://books.google.es/books?id=_vp1mwUwOmoC&lpg=PA644&dq=&pg=PA644#v=onepage&q&f=true), Ediciones Rialp, pp. 644, ISBN 9788432121005, .

[30] Diego-Fernández Sotelo, Rafael (1987) (in Spanish). *Las capitulaciones colombinas* (http://books.google.es/books?id=4ouNv6ZaRZQC&lpg=PA103&dq=&pg=PA139#v=onepage&q=&f=true). El Colegio de Michoacán A.C.. p. 139. ISBN 9789687230306. .

[31] Diego-Fernández Sotelo, Rafael (1987) (in Spanish). *Las capitulaciones colombinas* (http://books.google.es/books?id=4ouNv6ZaRZQC&lpg=PA103&dq=&pg=PA145#v=onepage&q=&f=true). El Colegio de Michoacán A.C.. pp. 143–145. ISBN 9789687230306. .

[32] Diego-Fernández Sotelo, Rafael (1987) (in Spanish). *Las capitulaciones colombinas* (http://books.google.es/books?id=4ouNv6ZaRZQC&lpg=PA103&dq=&pg=PA146#v=onepage&q=&f=true). El Colegio de Michoacán A.C.. p. 139. ISBN 9789687230306. .

[33] Diego-Fernández Sotelo, Rafael (1987) (in Spanish). *Las capitulaciones colombinas* (http://books.google.es/books?id=4ouNv6ZaRZQC&lpg=PA103&dq=&pg=PA149#v=onepage&q=&f=true). El Colegio de Michoacán A.C.. pp. 147–149. ISBN 9789687230306. .

[34] Sibaja Chacón, Luis Fernando (2006) (in Spanish). *El cuarto viaje de Cristóbal Colón y los orígenes de la provincia de Costa Rica* (http://books.google.es/books?id=q61T1fs_K4wC&lpg=PA120&dq=&pg=PA117#v=onepage&q=&f=false). EUNED. p. 117. ISBN 9789968314886. .

[35] Arranz Márquez, Luis (1982) (in Spanish). *Don Diego Colón, almirante, virrey y gobernador de las Indias* (http://books.google.es/books?id=kRhygUNmg4UC&lpg=PA102&dq=&pg=PA89#v=onepage&q=&f=true). CSIC. pp. 89–90. ISBN 9788400051563. .

[36] Arranz Márquez, Luis (1982) (in Spanish). *Don Diego Colón, almirante, virrey y gobernador de las Indias* (http://books.google.es/books?id=kRhygUNmg4UC&lpg=PA102&dq=&pg=PA97#v=onepage&q=&f=true). CSIC. p. 97. ISBN 9788400051563. .

[37] Rialp, Ediciones, S.A. (1992) (in Spanish). *Historia general de España y América* (http://books.google.es/books?id=4DWBNjs8iwEC&lpg=PA180&dq=&pg=PA195#v=onepage&q=&f=true). **10**. Ediciones Rialp. p. 195. ISBN 9788432121029. .

[38] Arranz Márquez, Luis (1982) (in Spanish). *Don Diego Colón, almirante, virrey y gobernador de las Indias* (http://books.google.es/books?id=kRhygUNmg4UC&lpg=PA102&dq=&pg=PA101#v=onepage&q=&f=true). CSIC. p. 101. ISBN 9788400051563. .

[39] Sibaja Chacón, Luis Fernando (2006) (in Spanish). *El cuarto viaje de Cristóbal Colón y los orígenes de la provincia de Costa Rica* (http://books.google.es/books?id=q61T1fs_K4wC&lpg=PA39&dq=&pg=PA39#v=onepage&q=&f=false). EUNED. p. 39. ISBN 9789968314886. .

[40] Rialp, Ediciones, S.A. (1992) (in Spanish). *Historia general de España y América* (http://books.google.es/books?id=4DWBNjs8iwEC&lpg=PA174&dq=&pg=PA174#v=onepage&q=&f=true). **10**. Ediciones Rialp. p. 174. ISBN 9788432121029. .

[41] Rialp, Ediciones, S.A. (1992) (in Spanish). *Historia general de España y América* (http://books.google.es/books?id=4DWBNjs8iwEC&lpg=PA183&dq=&pg=PA186#v=onepage&q=&f=true). **10**. Ediciones Rialp. p. 186. ISBN 9788432121029. .

[42] Rialp, Ediciones, S.A. (1992) (in Spanish). *Historia general de España y América* (http://books.google.es/books?id=4DWBNjs8iwEC&lpg=PA180&dq=&pg=PA196#v=onepage&q=&f=true). **10**. Ediciones Rialp. p. 195. ISBN 9788432121029. .

[43] Sibaja Chacón, Luis Fernando (2006) (in Spanish). *El cuarto viaje de Cristóbal Colón y los orígenes de la provincia de Costa Rica* (http://books.google.es/books?id=q61T1fs_K4wC&lpg=PA120&pg=PA36#v=onepage&q=&f=false). EUNED. p. 36. ISBN 9789968314886. .

[44] Rialp, Ediciones, S.A. (1992) (in Spanish). *Historia general de España y América* (http://books.google.es/books?id=4DWBNjs8iwEC&lpg=PA174&dq=&pg=PA197#v=onepage&q=&f=true). **10**. Ediciones Rialp. p. 197. ISBN 9788432121029. .

[45] Carrera Damas, Germán (1999) (in Spanish). *Historia general de América Latina* (http://books.google.es/books?id=Bq2DZpHvPkkC&lpg=PA457&dq=&pg=PA457#v=onepage&q&f=true). UNESCO. p. 457. ISBN 9789233031517. .

[46] Mena García, María del Carmen (1992) (in Spanish). *Pedrarias Dávila* (http://books.google.es/books?id=o9qselEBMysC&lpg=PA27&dq=&pg=PA29#v=onepage&q=&f=false). Universidad de Sevilla. p. 29. ISBN 9788474058345. .

[47] Sibaja Chacón, Luis Fernando (2006) (in Spanish). *El cuarto viaje de Cristóbal Colón y los orígenes de la provincia de Costa Rica* (http://books.google.es/books?id=q61T1fs_K4wC&lpg=PA58&pg=PA50#v=onepage&q&f=false). EUNED. p. 50. ISBN 9789968314886. .

[48] Sibaja Chacón, Luis Fernando (2006) (in Spanish). *El cuarto viaje de Cristóbal Colón y los orígenes de la provincia de Costa Rica* (http://books.google.es/books?id=q61T1fs_K4wC&lpg=PA58&pg=PA57#v=onepage&q&f=false). EUNED. pp. 55–59. ISBN 9789968314886. .

[49] Sibaja Chacón, Luis Fernando (2006) (in Spanish). *El cuarto viaje de Cristóbal Colón y los orígenes de la provincia de Costa Rica* (http://books.google.es/books?id=q61T1fs_K4wC&lpg=PA58&pg=PA32#v=onepage&q=&f=false). EUNED. p. 32. ISBN 9789968314886. .

[50] Rialp, Ediciones, S.A. (1992) (in Spanish). *Historia general de España y América* (http://books.google.es/books?id=4DWBNjs8iwEC&lpg=PA165&dq=&pg=PA165#v=onepage&q&f=false). **10**. Ediciones Rialp. p. 165. ISBN 9788432121029. .

[51] Sibaja Chacón, Luis Fernando (2006) (in Spanish). *El cuarto viaje de Cristóbal Colón y los orígenes de la provincia de Costa Rica* (http://books.google.es/books?id=q61T1fs_K4wC&lpg=PA58&pg=PA37#v=onepage&q&f=false). EUNED. pp. 36–37. ISBN 9789968314886. .

[52] Colón de Carvajal, Anunciada; Chocano Higueras, Guadalupe (1992) (in Spanish). *Cristóbal Colón: incógnitas de su muerte 1506-1902* (http://books.google.es/books?id=jBTF3fZowzUC&lpg=PA38&dq=&pg=PA40#v=onepage&q&f=false). CSIC. p. 40. ISBN 9788400073053. .

[53] Carrera Damas, Germán (1999) (in Spanish). *Historia general de América Latina* (http://books.google.es/books?id=Bq2DZpHvPkkC&lpg=PA457&dq=&pg=PA458#v=onepage&q&f=true). UNESCO. p. 458. ISBN 9789233031517. .

[54] Quoted in Fernand Braudel, *The Wheels of Commerce*, vol. II of *Civilization and Capitalism 15th–18th Centuryu* 1979:171.

[55] Braudel 1984

[56] Archer 2002, p. 251

[57] Tellier, Luc-Normand (2009), *Urban world history: an economic and geographical perspective* (http://books.google.com/books?id=cXuCjDbxC1YC), PUQ, p. 308, ISBN 2-760-51588-5, , Extract of page 308 (http://books.google.com/books?id=cXuCjDbxC1YC&pg=PA308)

[58] Quoted by Braudel 1984

[59] Elliott, 'Decline of Spain', pp. 56–57. Paul Kennedy points out that the very reliance on such a narrow tax base was a major problem for Spanish finances in the long term. See Kennedy, *Rise and Fall*, p. 68. (http://everything2.com/e2node/Decline%20of%20Spain)

[60] Chapter 15: A History of Spain and Portugal (http://libro.uca.edu/payne1/payne15.htm), Stanley G. Payne

[61] For a general account, see Kennedy, *Rise and Fall*, pp. 40–93.

[62] Brown & Elliott 1980, p. 190

[63] Cross and Crescent (http://mek.oszk.hu/01900/01911/html/index8.html)

[64] *When Europeans were slaves: Research suggests white slavery was much more common than previously believed* (http://researchnews.osu.edu/archive/whtslav.htm), , *Ohio State Research Communications* (Ohio State University), March 8, 2004, , retrieved 2008-10-08

[65] (http://books.google.es/books?id=sLD6aT7QfYgC&lpg=PA122&dq=&pg=PA122#v=onepage&q=&f=true)

[66] Conquest in the Americas (http://web.archive.org/web/20091028035130/http://encarta.msn.com/encyclopedia_761575057_13/Spain.html)

[67] Axtell, James (September/October 1991), "The Columbian Mosaic in Colonial America" (http://www.millersville.edu/~columbus/data/art/AXTELL01.ART), *Humanities* **12** (5): 12–18, , retrieved 2008-10-08

[68] http://www.unrc.edu.ar/publicar/tefros/revista/v6n1i08/paquetes/gascon.pdf

[69] "1512–1513: Laws of Burgos" (http://faculty.smu.edu/bakewell/BAKEWELL/texts/burgoslaws.html), *Colonial Latin America*, Peter Bakewell, 1998, , retrieved 2008-10-08

[70] Payne, Stanley G. (1973), "The Seventeenth-Century Decline" (http://libro.uca.edu/payne1/payne15.htm), *A History of Spain and Portugal*, **1**, Madison, Wisconsin: University of Wisconsin Press, , retrieved 2008-10-08

[71] Albareda Salvadó, Joaquim (2010). *La Guerra de Sucesión de España (1700–1714)* (http://books.google.es/books?id=3optq8WlRIMC&pg=PA240&dq=&hl=es&ei=QgTgTam5Ocy08QPD2amCCg&sa=X&oi=book_result&ct=result&resnum=1&ved=0CCsQ6AEwAA#v=onepage&q&f=true). Editorial Critica. pp. 239–241. .

[72] An early bandeira in 1628, (led by Antônio Raposo Tavares), composed of 2,000 allied Indians, 900 Mamluks (Mestizos) and 69 white Paulistanos, to find precious metals and stones and/or to capture Indians for slavery. This expedition alone was responsible for the destruction of most of the Jesuit missions of Spanish Guairá and the enslavement of 60,000 indigenous people. In response the missions that followed were heavily fortified.

Bibliography

- Anderson, James Maxwell (2000), *The History of Portugal*, Westport, Connecticut: Greenwood, ISBN 978-0313311062.
- Archer, Christon et al. (2002), *World History of Warfare*, Lincoln: University of Nebraska Press, ISBN 978-0803244238.
- Brown, Jonathan; Elliott, John Huxtable (1980), *A Palace for a King. The Buen Retiro and the Court of Philip IV*, New Haven: Yale University Press, ISBN 978-0300025071.
- Kamen, Henry (2003), *Empire: How Spain Became a World Power, 1492–1763*, New York: HarperCollins, ISBN 0-06-093264-3.
- Lach, Donald F.; Van Kley, Edwin J. (1994), *Asia in the Making of Europe*, Chicago: University of Chicago Press, ISBN 978-0226467344.
- Lockhart, James; Schwartz, Stuart B. (1983), *Early Latin America: A History of Colonial Spanish America and Brazil*, Cambridge: Cambridge University Press, ISBN 978-0521299299.

Further reading

- Armstrong, Edward (1902). *The emperor Charles V*. New York: The Macmillan Company
- Black, Jeremy (1996). *The Cambridge illustrated atlas of warfare: Renaissance to revolution*. Cambridge: Cambridge University Press. ISBN 0-521-47033-1
- Braudel, Fernand (1972). *The Mediterranean and the Mediterranean World in the Age of Philip II*, trans. Siân Reynolds. New York: Harper & Row. ISBN 0-06-090566-2
- Fernand Braudel, *The Perspective of the World* (part iii of *Civilization and Capitalism*) 1979, translated 1985.
- Brown, Jonathan (1998). *Painting in Spain : 1500–1700*. New Haven: Yale University Press. ISBN 0-300-06472-1
- Dominguez Ortiz, Antonio (1971). *The golden age of Spain, 1516–1659*. Oxford: Oxford University Press. ISBN 0-297-00405-0
- Edwards, John (2000). *The Spain of the Catholic Monarchs, 1474–1520*. New York: Blackwell. ISBN 0-631-16165-1
- Harman, Alec (1969). *Late Renaissance and Baroque music*. New York: Schocken Books.
- Kamen, Henry (1998). *Philip of Spain*. New Haven and London: Yale University Press. ISBN 0-300-07800-5
- Kamen, Henry (2005). *Spain 1469–1714. A Society of Conflict* (3rd ed.) London and New York: Pearson Longman. ISBN 0-582-78464-6
- Parker, Geoffrey (1997). *The Thirty Years' War* (2nd ed.). New York: Routledge. ISBN 0-415-12883-8
- Parker, Geoffrey (1972). *The Army of flanders and the Spanish road, 1567–1659; the logistics of Spanish victory and defeat in the Low Countries' Wars.*. Cambridge: Cambridge University Press. ISBN 0-521-08462-8
- Parker, Geoffrey (1977). *The Dutch revolt*. Cambridge: Cambridge University Press. ISBN 0-8014-1136-X
- Parker, Geoffrey (1978). *Philip II*. Boston: Little, Brown. ISBN 0-316-69080-5
- Parker, Geoffrey (1997). *The general crisis of the seventeenth century*. New York: Routledge. ISBN 0-415-16518-0
- Ramsey, John Fraser (1973) *Spain: the rise of the first world power*. University of Alabama Press. ISBN 0817357041 9780817357047
- Stradling, R. A. (1988). *Philip IV and the government of Spain*. Cambridge: Cambridge University Press. ISBN 0-521-32333-9
- Thomas, Hugh (2004). *Rivers of Gold: The Rise of the Spanish Empire 1490–1522* Weidenfeld & Nicolson. ISBN 0-297-64563-3
- Thomas, Hugh (1997). *The Slave Trade; The History of the Atlantic Slave Trade 1440–1870*. London: Papermac. ISBN 0-333-73147-6
- Various (1983). *Historia de la literatura espanola*. Barcelona: Editorial Ariel

- Wright, Esmond, ed. (1984). *History of the World, Part II: The last five hundred years* (3rd ed.). New York: Hamlyn Publishing. ISBN 0-517-43644-2.

External links

- Library of Iberian Resources Online, Stanley G Payne *A History of Spain and Portugal* vol 1 Ch 13 "The Spanish Empire" (http://libro.uca.edu/payne1/spainport1.htm)
- The Mestizo-Mexicano-Indian History in the USA (http://www.sonic.net/~doretk/ArchiveARCHIVE/NATIVE AMERICAN/TheMestizo-Mexicano-Indi.html)
- Documentary Film, *Villa de Albuquerque* (http://www.knmetv.org/programming/villadeabq.php3/)
- The last Spanish colonies (http://www.webcitation.org/query?url=http://es.geocities.com/coloniasesp&date=2009-10-25+03:17:15) (**Spanish**)
- Francisco José Calderón Vázquez (2008) (in Spanish), *Fronteras, identidad, conflicto e interacción. Los Presidios Españoles en el Norte Africano* (http://www.eumed.net/libros/2008c/433/), ISBN 978-84-691-6786-1

Mongomo

Mongomo	
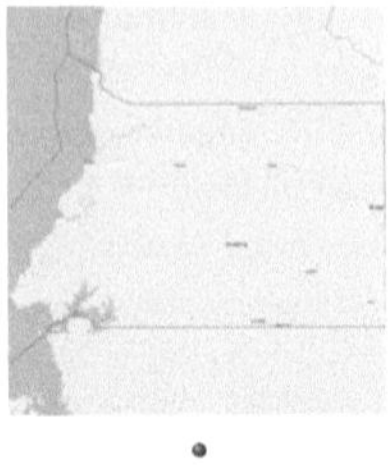 Mongomo Location in Equatorial Guinea	
Coordinates: 1°38′43.34″N 11°19′00.60″E	
Country	Equatorial Guinea
Province	Wele-Nzas

Mongomo is a city in Wele-Nzas Province in mainland Equatorial Guinea.

From Mongomo originate the nation's first president Francisco Macías Nguema and of the current President Teodoro Obiang Nguema Mbasogo, hence the name the **clan of Mongomo** to define those belonging to his ruling clique.

The area is home to Mongomo Provincial Hospital.[1]

The Mongomo Hotel is located in Kos Ete, Mongomo.[2]

Mongomo's football club is Deportivo Mongomo, active in the Equatoguinean Premier League.

References

[1] http://www.cababstractsplus.org/abstracts/Abstract.aspx?AcNo=20053014090
[2] http://www.mbendi.com/travel/venu/21/f790.htm

United States Geological Survey

United States Geological Survey	
Seal	
Official identifier	
Agency overview	
Formed	March 3, 1879
Headquarters	Reston, Virginia
Employees	8,670 (2009)
Annual budget	$1.1 billion (FY2010)[1]
Agency executive	Marcia McNutt, [2] Director
Parent agency	Department of the Interior
Website	
usgs.gov [3]	

The **United States Geological Survey** (**USGS**) is a scientific agency of the United States government. The scientists of the USGS study the landscape of the United States, its natural resources, and the natural hazards that threaten it. The organization has four major science disciplines, concerning biology, geography, geology, and hydrology. The USGS is a fact-finding research organization with no regulatory responsibility.

A bureau of the United States Department of the Interior, it is that department's sole scientific agency. The USGS employs approximately 8,670 people[4] and is headquartered in Reston, Virginia. The USGS also has major offices in Lakewood, Colorado (Denver Federal Center), and Menlo Park, California.

The motto of the USGS is "*Science for a changing world.*"[5]

History

Prompted by a report from the National Academy of Sciences the USGS was created by an act of Congress on March 3, 1879. It was charged with the "classification of the public lands, and examination of the geological structure, mineral resources, and products of the national domain." This task was driven by the need to inventory the vast lands added to the United States by the Louisiana Purchase in 1803 and the Mexican-American War in 1848.

Clarence King, the first director of USGS, assembled the new organization from disparate regional survey agencies. After a short tenure, King was succeeded in the director's chair by John Wesley Powell.

List of USGS Directors

Clarence King, founder of the USGS

- 1879–1881 Clarence King
- 1881–1894 John Wesley Powell
- 1894–1907 Charles Doolittle Walcott
- 1907–1930 George Otis Smith
- 1930–1943 Walter Curran Mendenhall
- 1943–1956 William Embry Wrather
- 1956–1965 Thomas Brennan Nolan
- 1965–1971 William Thomas Pecora
- 1971–1978 Vincent Ellis McKelvey
- 1978–1981 Henry William Menard
- 1981–1993 Dallas Lynn Peck
- 1994–1997 Gordon P. Eaton
- 1998–2005 Charles G. Groat
- 2006-2009 Mark Myers
- 2009- Marcia McNutt

Programs

The USGS is divided into seven Mission Areas[7] with specific programs including:

The USGS headquarters in Reston, VA

USGS gauging station 03221000 [6] on the Scioto River below O'Shaughnessy Dam near Dublin, Ohio

- Earthquake Hazards Program [8] monitors earthquake activity worldwide. The National Earthquake Information Center (NEIC) in Golden, Colorado on the campus of the Colorado School of Mines detects the location and magnitude of global earthquakes. The USGS also runs or supports several regional monitoring networks in the U.S. under the umbrella of the Advanced National Seismic System [9] (ANSS). The USGS informs authorities, emergency responders, the media, and the public, both domestic and worldwide, about significant earthquakes. It also maintains long-term archives of earthquake data for scientific and engineering research. It also conducts and supports research on long-term seismic hazards.
- As of 2005, the agency is working to create a National Volcano Early Warning System by improving the instrumentation monitoring the 169 volcanoes in U.S. territory and by establishing methods for measuring the relative threats posed at each site.
- The USGS National Geomagnetism Program monitors the magnetic field at magnetic observatories and distributes magnetometer data in real time.
- The USGS collaborates with Canadian and Mexican government scientists, along with the Commission for Environmental Cooperation, to produce the North American Environmental Atlas, which is used to depict and track environmental issues for a continental perspective.
- The USGS operates the streamgaging network for the United States, with over 7400 streamgages. Real-time streamflow data [10] are available online.
- National Climate Change and Wildlife Science Center (NCCWSC) [11] implements partner-driven science to improve understanding of past and present land use change, develops relevant climate and land use forecasts, and identifies lands, resources, and communities that are most vulnerable to adverse impacts of change from the local to global scale.[12]
- Since 1962, the Astrogeology Research Program has been involved in global, lunar and planetary exploration and mapping.
- In collaboration with Stanford University, the USGS also operates the USGS-Stanford Ion Microprobe Laboratory [13], a world-class[14] analytical facility for U-(Th)-Pb geochronology and trace element analyses of minerals and other earth materials.

- USGS operates a number of water related programs, notably the National Streamflow Information Program [15] and National Water-Quality Assessment Program [16]. USGS Water data is publicly available from their National Water Information System [17] database.
- The USGS also operates the National Wildlife Health Center, whose mission is "to serve the nation and its natural resources by providing sound science and technical support, and to disseminate information to promote science-based decisions affecting wildlife and ecosystem health. The NWHC provides information, technical assistance, research, education, and leadership on national and international wildlife health issues."[18] It is the agency primarily responsible for surveillance of wild-animal H5N1 avian influenza outbreaks in the United States. The USGS also runs 17 biological research centers in the United States, including the Patuxent Wildlife Research Center.

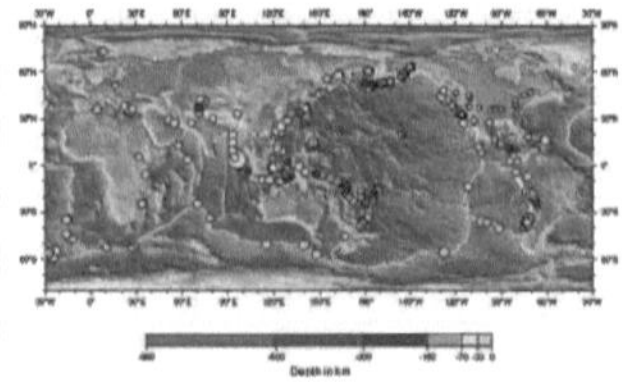

Recent earthquakes from the last 8-30 days around the world.

Topographic Mapping

The United States Geological Survey (USGS), a civilian federal agency, produces several national series of topographic maps which vary in scale and extent, with some wide gaps in coverage, notably the complete absence of 1:50,000 scale topographic maps or their equivalent. The largest (both in terms of scale and quantity) and best-known topographic series is the 7.5-minute, 1:24,000 scale, quadrangle, a non-metric scale virtually unique to the United States. Each of these maps covers an area bounded by two lines of latitude and two lines of longitude spaced 7.5 minutes apart. Nearly 57,000 individual maps in this series cover the 48 contiguous states, Hawaii, U. S. territories, and areas of Alaska near Anchorage, Fairbanks, and Prudhoe Bay. The area covered by each map varies with the latitude of its represented location due to convergence of the meridians. At lower latitudes, near 30° north, a 7.5-minute quadrangle contains an area of about 64 square miles (166 km^2). At 49° north latitude, 49 square miles (127 km^2) are contained within a quadrangle of that size. As a unique non-metric map scale, the 1:24,000 scale naturally requires a separate and specialized romer scale for plotting map positions.[19] In recent years, budget constraints have forced the USGS to rely on donations of time by civilian volunteers in an attempt to update its 7.5-minute topographic map series, and USGS stated outright in 2000 that the program was to be phased out in favor of *The National Map*[20] (not to be confused with the National Atlas of the United States produced by the Department of the Interior, one of whose bureaus is USGS).

An older series of maps, the 15-minute series, was once used to map the contiguous 48 states at a scale of 1:62,500, but was discontinued some time ago for maps covering the continental U.S. Each map was bounded by two parallels and two meridians spaced 15 minutes apart - the same area covered by four maps in the 7.5-minute series. The 15-minute series, at a scale of 1:63,360 (one inch representing one mile), remains the primary topographic quadrangle for the state of Alaska (and only for that particular state). Nearly 3,000 maps cover 97% of the state.[19] The U.S.A. remains virtually the only developed country in the world without a standardized civilian topographic map series in the standard 1:25,000 or 1:50,000 metric scales, making coordination difficult in border regions (the U.S. military does issue 1:50,000 scale topo maps of the continental U.S., though only for use by members of its defense forces).

The next-smallest topographic series, in terms of scale, is the 1:100,000 series. These maps are bounded by two lines of longitude and two lines of latitude. However, in this series, the lines of latitude are spaced 30 minutes apart and the lines of longitude are spaced 60 minutes, which is the source of another name for these maps; the 30 x 60-minute quadrangle series. Each of these quadrangles covers the area contained within 32 maps in the 7.5-minute series. The 1:100,000 scale series is unusual in that it employs the Metric system primarily. One centimeter on the map represents one kilometer of distance on the ground. Contour intervals, spot elevations, and horizontal distances are also specified in meters.

The final regular quadrangle series produced by the USGS is the 1:250,000 scale topographic series. Each of these quadrangles in the conterminous United States measures 1 degree of latitude by 2 degrees of longitude. This series was produced by the U.S. Army Map Service in the 1950s, prior to the maps in the larger-scale series, and consists of 489 sheets, each covering an area ranging from 8218 square miles (21285 km^2) at 30° north to 6222 square miles (16115 km^2) at 49° north.[19] Hawaii is mapped at this scale in quadrangles measuring 1° by 1°.

USGS topographic quadrangle maps are marked with grid lines and tics around the map collar which make it possible to identify locations on the map by several methods, including the graticule measurements of longitude and latitude, the township and section method within the Public Land Survey System, and cartesian coordinates in both the State Plane Coordinate System and the Universal Transverse Mercator coordinate system.

Other specialty maps have been produced by the USGS at a variety of scales. These include county maps, maps of special interest areas, such as the national parks, and areas of scientific interest.

A number of Internet sites have made these maps available on the web for affordable commercial and professional use. Because works of the U.S. Government are in the public domain, it is also possible to find many of these maps for free at various locations on the Internet. Georeferenced map images are available from the USGS as digital raster graphics (DRGs) in addition to digital data sets based on USGS maps, notably Digital Line Graphs (DLGs) and digital elevation models (DEMs).

USGS publications

USGS publishes many series of maps and reports, including:

Biological Science Report (BSR)

Record significant scientific interpretations and findings, usually of lasting scientific interest, addressing a wide variety of topics relevant to Biological Resources Discipline (BRD) investigations and research. May include extensive data or theoretical analyses. Reports published by the U.S. Biological Survey and later by the U.S. Geological Survey. The report series began in 1995 and continued through 2003.

Bulletin (B)

Significant data and interpretations of lasting scientific interest but generally narrower in scope than professional papers. Results of resource studies, geologic or topographic studies, and collections of short papers on related topics.

Circular (CIR/C)

A wide variety of topics covered concisely and clearly to provide a synthesis of understanding about processes, geographic areas, issues, or USGS programs. The Circular should be aimed at enhancing knowledge and understanding among general audiences, decision makers, university students, and scientists in related fields.

Circum-Pacific Map (CP)

Multicolor equal-area maps at scales of 1:10,000,000 for the Northwest, Northeast, Southwest, Southeast quadrants of the Pacific and the Arctic and Antarctic regions, and of 1:17,000,000 for the whole Pacific Basin. The series consists of base, geographic, geodynamic, plate-tectonic, geologic, tectonic, mineral-resources, and energy-resources maps, as well as other miscellaneous maps.

Coal Investigations (COAL/C-) Map

Origin, character, and resource potential of coal deposits shown by geologic maps, structure contours, cross sections, columnar sections, and measured coal sections, where appropriate. Text on same sheet or in an accompanying pamphlet.

Data Series (DS)

The Data Series is intended for release of basic data sets, databases, and multimedia or motion graphics. This series can be used for videos, computer programs, and collections of digital photographs.

Folios of the Geologic Atlas (GF)

Quadrangles named from a city, town, or prominent natural feature within the area covered. They include maps showing the topography, geology, underground structure and mineral deposits of the area and several pages of descriptive text and illustrations. May include maps of oil and gas and artesian water. Precursor to Geologic Quadrangles.

General Interest Publication (GIP)

A wide variety of topics covered concisely and clearly in a variety of formats. Focus is on USGS programs, projects, and services and general scientific information of public interest. The series covers a broad range of topics in a variety of media, including pamphlets, postcards, posters, videos, teacher kits, CD/DVDs, bookmarks, and interactive and motion graphics. Previously called "General Interest Publications".

Geologic Quadrangle (GQ) Map

Detailed geologic maps depicting areas of special importance to the solution of geologic problems. May portray bedrock or surficial units, or both. May include brief texts, structure sections, and columnar sections. 71/2- or 15-minute quadrangles printed in multicolor on topographic bases that meet National Map Accuracy standards.

Geophysical Investigations (GP) Map

Chiefly the results of aeromagnetic and (or) gravity surveys shown by contours. Area depicted may range in size from a few square miles to an entire country. Single or multiple sheets.

Hydrologic Investigations Atlas (HA)

A wide range of hydrologic and hydrogeologic data of regional and national interest, such as streamflow, ground water, water quality, and extent of flooding. Various scales. Single or multiple sheets.

Land Use and Land Cover (L) Map

Various categories of land use and cover, both artificial and natural, for use by geographers, land-use planners, and others. Planimetric maps at scales of 1:250,000 or 1:100,000 on a single sheet.

Mineral Investigations Resource (MR) Map

Information on mineral occurrences, mineral resources, mines and prospects, commodities, and target areas of possible resources **other than** coal, petroleum, or natural gas. Small scale (1:250,000 or smaller).

Miscellaneous Field Studies (MF) Map

Rapidly prepared, low-budget maps in a broad range of presentations in terms of portrayal, completeness, interpretations, draftsmanship, scale, and area coverage. Single or multiple sheets.

Miscellaneous Investigations/ Geologic Investigations (I) Series

High-quality maps and charts of varied subject matter such as bathymetry, geology, hydrogeology, landforms, land-use classification, vegetation, and others including maps of planets, the Moon, and other satellites. Various scales. Topographic or planimetric bases; regular or irregular areas. May include a text printed as an accompanying pamphlet.

Oil and Gas Investigations (OC) Chart

Information about known or possible petroleum resources, presented as logs, correlation diagrams, graphs, and tables, but ordinarily not as maps. Single or multiple sheets. Text printed on same sheet or in an accompanying pamphlet.

Oil and Gas Investigations (OM) Map

Apply particularly to areas of known or possible petroleum resources. Typically include cross sections, columnar sections, structure contours, correlation diagrams, and information on wells drilled for oil and gas. Single or multiple sheets. Text usually on map sheet but sometimes printed as an accompanying pamphlet.

Open-File Report (OFR/OF)

Interpretive information that needs to be released immediately; maps and reports (and their supporting data) that need to be released as supporting documentation because they are referenced, discussed, or interpreted in another information product; preliminary findings (pending a final map or report); interim computer programs and user guides; bibliographies.

Professional Paper (PP)

Premier series of the USGS. Comprehensive reports of wide and lasting interest and scientific importance, characterized by thoroughness of study and breadth of scientific or geographic coverage. The series may include collections of related papers addressing different aspects of a single scientific topic, either issued together under one cover or separately as chapters.

Water-Resources Investigations Report (WRIR/WRI)

Hydrologic information, mainly of local interest, intended for quick release. Book or map format. Varied scales.

Water-Supply Paper (WSP)

Reports on all aspects of hydrology, including quality, recoverability, and use of water resources; statistical reports on streamflow, floods, groundwater levels, and water quality; and collections of short papers on related topics.

A complete listing of descriptions of USGS Series is available at [21] (accessed 11/25/08)

Locating USGS Publications USGS publication are available for purchase at USGS Publications Warehouse [22].

Many USGS publications are now available online:

- USGS Publications Warehouse [23], Search for online documents
- Open-File reports online [24], accessed 11/25/08
- Mineral Resources Program [25], Online Publications and Data
- Central Mineral Resources Team [26], selected maps and publications online
- National Strong-Motion Project [27] Reports and software online.

Many older USGS publications have been scanned and digitized by such services as Google Books. An online search will quickly reveal if a digital version is available. All USGS publications are public domain.

Controversy

In December 2006, it was announced that the rules for the publication of USGS research were being revised. Employees were informed that USGS leadership and communications staff should be notified "of findings or data that may be especially newsworthy, have an impact on government policy, or contradict previous public understanding to ensure that proper officials are notified and that communication strategies are developed."[28] [29]

The revision was claimed not to change existing rules, but rather to emphasize the importance of maintaining the scientific integrity of the agency's work by requiring scientists to accept comments from the public and follow administrative policies. However, scientists have questioned whether this revision is likely to facilitate censorship of their work, as has been alleged by critics to have occurred in some federal agencies under the administration of United States President George W. Bush.[28] [30]

According to the authors of this policy, USGS information is given to the public after it has been through a peer review and approval process. USGS leadership and communications staff are kept informed of relevant scientific findings so they can manage the flow of information to decision-makers, who use this information to make resource-management choices. Policy makers have said these principles and practices will bolster the USGS's scientific objectivity and reputation.[28] [29] [31] [32]

See also

- Core research center
- Geographic Names Information System
- Maps of the United States
- National LIDAR Dataset - USA
- QuakeSim
- Timeline of environmental events
- Volcano Disaster Assistance Program
- Hawaiian Volcano Observatory

Notes

[1] "Secretary Salazar Applauds Senate's Confirmation of Dr. Marcia McNutt as Director of the U.S. Geological Survey" (http://www.doi.gov/news/09_News_Releases/102209.html) (Press release). U.S. Department of the Interior. October 22, 2009. . Retrieved 2009-11-07.

[2] Straub, Noelle (October 22, 2009). "Senate Confirms Nominees for Interior, DOE" (http://www.nytimes.com/gwire/2009/10/22/22greenwire-senate-confirms-nominees-for-interior-doe-23080.html). The New York Times. . Retrieved 2009-10-22.

[3] http://www.usgs.gov

[4] The Associated Press (October 23, 2009). "Monterey Aquarium's McNutt new USGS director" (http://www.google.com/hostednews/ap/article/ALeqM5h8x87AaAYRWeBc4YNW2FfTS7yLpgD9BH4EEO0). Google News. . Retrieved 2009-10-25.

[5] "USGS Visual Identity System" (http://www.usgs.gov/visual-id/outside_use.html). U.S. Geological Survey. July 27, 2006. . Retrieved 2008-12-29.

[6] http://waterdata.usgs.gov/oh/nwis/uv?03221000

[7] "Start with Science Web site" (http://www.usgs.gov/start_with_science/). . Retrieved 2011-07-20.

[8] http://earthquake.usgs.gov/

[9] http://earthquake.usgs.gov/research/monitoring/anss/

[10] http://water.usgs.gov/waterwatch/

[11] http://nccwsc.usgs.gov/index.shtml

[12] "NCCWSC Web ite" (http://nccwsc.usgs.gov/index.shtml). . Retrieved 2011-07-20.

[13] http://shrimprg.stanford.edu

[14] "SHRIMP-RG Bibliography as of 24, July 2007" (http://shrimprg.stanford.edu/bibliography/Ion_Probe_Bibliography7-07.htm). .

[15] http://water.usgs.gov/nsip

[16] http://water.usgs.gov/nawqa

[17] http://water.usgs.gov/data

[18] "National Wildlife Health Center" (http://www.nwhc.usgs.gov/). . Retrieved 2007-12-27.

[19] USGS Topographic Maps (http://topomaps.usgs.gov/) and "USGS Maps Booklet" (http://erg.usgs.gov/isb/pubs/booklets/usgsmaps/usgsmaps.html). .

[20] " The U.S. Geological Survey's Revision Program for 7.5-Minute Topographic Maps (http://thor-f5.er.usgs.gov/topomaps/revision_overview.pdf)", Larry Moore (USGS), December, 2000.

[21] http://infotrek.er.usgs.gov/pls/htmldb/f?p=127:13:1996281797770541

[22] http://pubs.er.usgs.gov/

[23] http://pubs.er.usgs.gov/usgspubs/index.jsp?view=adv

[24] http://pubs.usgs.gov/of/

[25] http://minerals.usgs.gov/products/index.html

[26] http://minerals.cr.usgs.gov/publications/index.html

[27] http://nsmp.wr.usgs.gov/publications/online_reports.html

[28] Lubick, February 7, 2007

[29] Eilperin, December 14, 2006

[30] Eilperin, December 14, 2006. Eilperin uses information without attribution, quoting the one person mentioned in Heilprin's earlier article: James Estes.

[31] Erickson, December 14, 2006

[32] Heilprin, December 13, 2006

References

- Eilperin, Juliet (December 14, 2006). "USGS Scientists Object To Stricter Review Rules" (http://www.washingtonpost.com/wp-dyn/content/article/2006/12/13/AR2006121301991.html). The Washington Post. Retrieved 2007-12-27.
- Erickson, Jim (December 14, 2006). "USGS to review scientists' studies prior to publication" (http://web.archive.org/web/20080108011139/http://www.rockymountainnews.com/drmn/local/article/0,1299,DRMN_15_5213051,00.html). Rocky Mountain News. Archived from the original (http://www.rockymountainnews.com/drmn/local/article/0,1299,DRMN_15_5213051,00.html) on 2008-01-08. Retrieved 2007-12-27.
- Heilprin, John (December 13, 2006). "USGS scientists worry about being muzzled" (http://www.msnbc.msn.com/id/16193443/). Associated Press. Retrieved 2007-12-27.
- Lubick, Naomi (February 7, 2007). "Heavier peer review policy for USGS" (http://pubs.acs.org/subscribe/journals/esthag-w/2007/feb/policy/nl_prpolicy.html). Environmental Science & Technology. Retrieved 2007-12-27.
- "500.5 - News Release and Media Relations Policy" (http://www.usgs.gov/usgs-manual/500/500-5.html). U.S. Geological Survey. August 19, 2002. Retrieved 2007-12-27.
- "502.4 - Fundamental Science Practices: Review, Approval, and Release of Information Products" (http://www.usgs.gov/usgs-manual/500/502-4.html). U.S. Geological Survey. September 20, 2006. Retrieved 2007-12-27.

External links

USGS sites

- USGS official site (http://www.usgs.gov/)
- Major USGS Mission Area sites: Water (http://water.usgs.gov/), Energy, Minerals, and Environmental Health (http://www.usgs.gov/resources_envirohealth/), Climate and Land Use Change (http://www.usgs.gov/climate_landuse/), Ecosystems (http://www.usgs.gov/ecosystems/), Natural Hazards (http://www.usgs.gov/natural_hazards/), Core Science Systems (http://www.usgs.gov/core_science_systems/)
- Science Quality and Integrity (http://www.usgs.gov/quality_integrity/)
- USGS Fundamental Science Practices (http://www.usgs.gov/fsp/)
- USGS Library (http://library.usgs.gov/)
- Science Topics (http://www.usgs.gov/science/) (thematic directory)
- Science in Your Backyard (http://www.usgs.gov/state/) (State index)
- USGS Circular 1050 (History of the USGS) (http://pubs.usgs.gov/circ/c1050/)
- USGS Geomagnetism Program (http://geomag.usgs.gov)
- Earthquake Hazards Program (http://earthquake.usgs.gov/) (current and historical earthquake information)
- Subscribe to automated earthquake messages (http://earthquake.usgs.gov/eqcenter/ens/)
- Volcano Information (http://volcanoes.usgs.gov/)
- Invasive Species (http://ecosystems.usgs.gov/invasive/)
- Land Remote Sensing (http://remotesensing.usgs.gov/index.php)
- USGS Realtime Streamflow Information (http://water.usgs.gov/waterwatch)
- USGS Water Data for the Nation (http://waterdata.usgs.gov/nwis)

Non-USGS related sites

- The Libre Map Project (http://libre.redjar.org/maps/) offers free, high-quality USGS DRG maps in TIFF format along with world files for use with your GIS software
- TerraServer-USA (http://www.terraserver-usa.com/) and Acme (http://mapper.acme.com/) host USGS topographic maps (and aerial photos on TerraServer-USA);
- Maptech (http://historical.maptech.com/) hosts historical USGS topos in the northeast U.S.

- Mapfinder Utility (http://www.usgsquads.com/index.php/map-indexes/mapfinder/) download USGS Topographic maps for free in Tiff format using Google Earth.
- The US Minerals Databrowser (http://mazamascience.com/Minerals/USGS/) provides interactive access to data visualizations based on data from USGS DS140: "Historical Statistics for Mineral Commodities"

Artisan

An **artisan** (from Italian: *artigiano*) is a skilled manual worker who makes items that may be functional or strictly decorative, including furniture, clothing, jewellery, household items, and tools. As an adjective (spelled "artisanal"), it has been used as a marketing buzz word to describe or imply an association with the crafting of hand made food products, such as bread, tofu, beverages and cheese.

A Cameroonian artisan at work

Artisan origins

Artisans were the dominant producers of consumer products prior to the Industrial Revolution. According to classical economics theory, the division of labour occurs with internal market development (Adam Smith). However, according to economist John Hicks, merchants and artisans originated as servants to the rulers, which occurred much earlier.

Medieval artisans

During the Middle Ages, the term "artisan" was applied to those who made things or provided services. It did not apply to unskilled labourers. Artisans were divided into two distinct groups: those who operated their own business, and those who did not. Those who owned their businesses were called masters, while the latter were the journeymen and apprentices. One misunderstanding many people have about this social group is that they picture them as "workers" in the modern sense: employed by someone. The most influential group among the artisans were the masters, the business owners. The owners enjoyed a higher social status in their communities.[1]

See also

- Artist
- Applied art
- Arts and Crafts movement
- Caste — Tarkhan

References

[1] History of Western Civilization, Boise State University http://history.boisestate.edu/westciv/medsoc/23.shtml

Gabon

Gabonese Republic
République Gabonaise

Motto: "Union, Travail, Justice"
(French for "Union, Work, Justice")

Anthem: *La Concorde*
The Concord

Capital (and largest city)		Libreville 0°23′N 9°27′E
Official language(s)		French
Vernacular languages		Fang, Myene
Demonym		Gabonese, Gabonaise
Government		Presidential republic
-	President	Ali Bongo Ondimba
-	Prime Minister	Paul Biyoghé Mba
Independence		
-	from France	August 17, 1960
Area		
-	Total	267,667 km^2 (76th) 103,347 sq mi
-	Water (%)	3.76%
Population		

-	2009 estimate	1,475,000[1] (150th)
-	Density	5.5/km² (216th) 14.3/sq mi
GDP (PPP)		2010 estimate
-	Total	$22.478 billion[2]
-	Per capita	$15,020[2]
GDP (nominal)		2010 estimate
-	Total	$13.056 billion[2]
-	Per capita	$8,724[2]
HDI (2010)		▲ 0.648[3] (medium) (93rd)
Currency		Central African CFA franc (XAF)
Time zone		WAT (UTC+1)
-	Summer (DST)	not observed (UTC+1)
Drives on the		right
ISO 3166 code		GA
Internet TLD		.ga
Calling code		241

Gabon (/gəˈbɒn/; French pronunciation: [gabɔ̃]), officially the **Gabonese Republic** (French: *République Gabonaise*) is a state in west central Africa sharing borders with Equatorial Guinea to the northwest, Cameroon to the north, and with the Republic of the Congo curving around the east and south. The Gulf of Guinea, an arm of the Atlantic Ocean is to the west. It covers a land area of nearly 270,000 km² and has an estimated population of 1,500,000. Its capital and largest city is Libreville.

Since its independence from France on August 17, 1960, Gabon has been ruled by three presidents. In the early 1990s, Gabon introduced a multi-party system and a new democratic constitution that allowed for a more transparent electoral process and reformed many governmental institutions. Gabon is also a non-permanent member of the United Nations Security Council for the 2010-2011 term. The small population density together with abundant natural resources and foreign private investment have helped make Gabon one of the most prosperous countries in Sub-Saharan Africa, with the fourth highest HDI[4] and the third highest GDP per capita (PPP) (after Equatorial Guinea and Botswana) in the region.

History

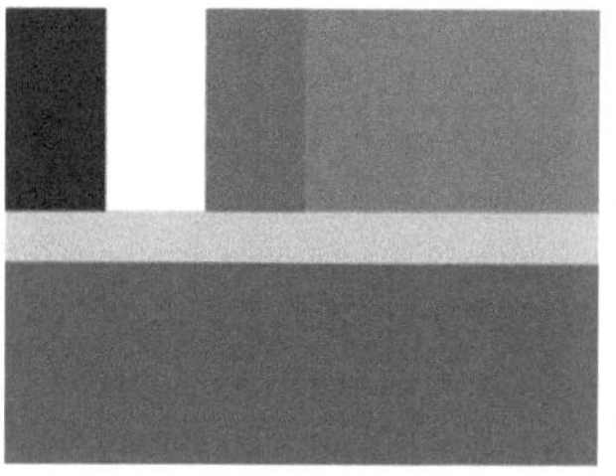

Flag of Gabon 1959-1960

The earliest inhabitants of the area were Pygmy peoples. They were largely replaced and absorbed by Bantu tribes as they migrated.

In the 15th century, the first Europeans arrived. The nation's present name originates from "Gabão", Portuguese for "cloak", which is roughly the shape of the estuary of the Komo River by Libreville. French explorer Pierre Savorgnan de Brazza led his first mission to the Gabon-Congo area in 1875. He founded the town of Franceville, and was later colonial governor. Several Bantu groups lived in the area that is now Gabon when France officially occupied it in 1885.

In 1910, Gabon became one of the four territories of French Equatorial Africa, a federation that survived until 1959. These territories became independent on August 17, 1960. The first president of Gabon, elected in 1961, was Léon M'ba, with Omar Bongo Ondimba as his vice president. French interests were decisive in selecting the future leadership in Gabon after Independence; French logging interests poured funds into the successful election campaign of M'ba, an *évolué* from the coastal region.

After M'ba's accession to power, the press was suppressed, political demonstrations banned, freedom of expression curtailed, other political parties gradually excluded from power and the Constitution changed along French lines to vest power in the Presidency, a post that M'ba assumed himself. However, when M'ba dissolved the National Assembly in January 1964 to institute one-party rule, an army coup sought to oust him from power and restore parliamentary democracy. The extent to which M'ba's dictatorial regime was synonymous with "French Interests" then became blatantly apparent when French paratroopers flew in within 24 hours to restore M'ba to power.

After a few days of fighting, the coup was over and the opposition imprisoned, despite widespread protests and riots. The French government was unperturbed by international condemnation of the intervention, and paratroops still remain in the Camp de Gaulle on the outskirts of Gabon's capital. When M'Ba died in 1967, Bongo replaced him as president.

In March 1968, Bongo declared Gabon a one-party state by dissolving the BDG and establishing a new party—the Parti Democratique Gabonais (PDG). He invited all Gabonese, regardless of previous political affiliation, to participate. Bongo sought to forge a single national movement in support of the government's development policies, using the PDG as a tool to submerge the regional and tribal rivalries that had divided Gabonese politics in the past. Bongo was elected President in February 1975; in April 1975, the position of vice president was abolished and replaced by the position of prime minister, who had no right to automatic succession. Bongo was re-elected President in both December 1979 and November 1986 to 7-year terms.[5]

Economic discontent and a desire for political liberalization provoked violent demonstrations and strikes by students and workers in early 1990. In response to grievances by workers, Bongo negotiated with them on a sector-by-sector basis, making significant wage concessions. In addition, he promised to open up the PDG and to organize a national political conference in March–April 1990 to discuss Gabon's future political system. The PDG and 74 political organizations attended the conference. Participants essentially divided into two loose coalitions, the ruling PDG and its allies, and the United Front of Opposition Associations and Parties, consisting of the breakaway Morena Fundamental and the Gabonese Progress Party.[5]

The April 1990 conference approved sweeping political reforms, including creation of a national Senate, decentralization of the budgetary process, freedom of assembly and press, and cancellation of an exit visa requirement. In an attempt to guide the political system's transformation to multiparty democracy, Bongo resigned as PDG chairman and created a transitional government headed by a new Prime Minister, Casimir Oye-Mba. The Gabonese Social Democratic Grouping (RSDG), as the resulting government was called, was smaller than the previous government and included representatives from several opposition parties in its cabinet. The RSDG drafted a provisional constitution in May 1990 that provided a basic bill of rights and an independent judiciary but retained strong executive powers for the president. After further review by a constitutional committee and the National Assembly, this document came into force in March 1991.[5]

Opposition to the PDG continued after the April 1990 conference, however, and in September 1990, two coup d'état attempts were uncovered and aborted. Despite anti-government demonstrations after the untimely death of an opposition leader, the first multiparty National Assembly elections in almost 30 years took place in September–October 1990, with the PDG garnering a large majority.[5]

Following President Omar Bongo's re-election in December 1993 with 51% of the vote, opposition candidates refused to validate the election results. Serious civil disturbances led to an agreement between the government and opposition factions to work toward a political settlement. These talks led to the Paris Accords in November 1994, under which several opposition figures were included in a government of national unity. This arrangement soon broke down, however, and the 1996 and 1997 legislative and municipal elections provided the background for renewed partisan politics. The PDG won a landslide victory in the legislative election, but several major cities, including Libreville, elected opposition mayors during the 1997 local election.[5]

Facing a divided opposition, President Omar Bongo coasted to easy re-election in December 1998, with large majorities of the vote. While Bongo's major opponents rejected the outcome as fraudulent, some international observers characterized the results as representative despite many perceived irregularities, and there were none of the civil disturbances that followed the 1993 election. Peaceful though flawed legislative elections held in 2001-2002, which were boycotted by a number of smaller opposition parties and were widely criticized for their administrative weaknesses, produced a National Assembly almost completely dominated by the PDG and allied independents. In November 2005, President Omar Bongo was elected for his sixth term. He won re-election easily, but opponents claim that the balloting process was marred by irregularities. There were some instances of violence following the announcement of Omar Bongo's win, but Gabon generally remained peaceful.[5]

National Assembly elections were held again in December 2006. Several seats contested because of voting irregularities were overturned by the Constitutional Court, but the subsequent run-off elections in early 2007 again yielded a PDG-controlled National Assembly.[5]

On June 8, 2009, President Omar Bongo died of cardiac arrest at a Spanish hospital in Barcelona, ushering in a new era in Gabonese politics. In accordance with the amended constitution, Rose Francine Rogombe, the President of the Senate, became Interim President on June 10, 2009. The first contested elections in Gabon's history that did not include Omar Bongo as a candidate were held on August 30, 2009 with 18 candidates for president. The lead-up to the elections saw some isolated protests, but no significant disturbances. Omar Bongo's son, ruling party leader Ali Bongo Ondimba, was formally declared the winner after a 3-week review by the Constitutional Court; his inauguration took place on October 16, 2009.[5]

The court's review had been prompted by claims of fraud by the many opposition candidates, with the initial announcement of election results sparking unprecedented violent protests in Port-Gentil, the country's second-largest city and a long-time bastion of opposition to PDG rule. The citizens of Port-Gentil took to the streets, and numerous shops and residences were burned, including the French Consulate and a local prison. Officially, only four deaths occurred during the riots, but opposition and local leaders claim many more. Gendarmes and the military were deployed to Port-Gentil to support the beleaguered police, and a curfew was in effect for more than 3 months.[5]

A partial legislative by-election was held in June 2010. A newly created coalition of parties, the Union Nationale (UN), participated for the first time. The UN is composed largely of PDG defectors who left the party after Omar Bongo's death. Of the five hotly contested seats, the PDG won three and the UN won two; both sides claimed victory.[5]

Government

Former President Omar Bongo Ondimba in 2004

Gabon is a republic with a presidential form of government under the 1961 constitution (revised in 1975, rewritten in 1991, and revised in 2003). The president is elected by universal suffrage for a 7-year term; a 2003 constitutional amendment removed presidential term limits and facilitated a presidency for life. The president can appoint and dismiss the prime minister, the cabinet, and judges of the independent Supreme Court. The president also has other strong powers, such as authority to dissolve the National Assembly, declare a state of siege, delay legislation, and conduct referenda.[5]

The country has a bicameral legislature with a National Assembly and Senate. The National Assembly has 120 deputies who are popularly elected for a 5-year term. The Senate is composed of 102 members who are elected by municipal councils and regional assemblies and serve for 6 years. The Senate was created in the 1990-1991 constitutional re-write, although it was not brought into being until after the 1997 local elections. The President of the Senate is next in succession to the President.[5]

In 1990 the government made major changes to Gabon's political system. A transitional constitution was drafted in May 1990 as an outgrowth of the national political conference in March–April and later revised by a constitutional committee. Among its provisions were a Western-style bill of rights, creation of a National Council of Democracy to oversee the guarantee of those rights, a governmental advisory board on economic and social issues, and an independent judiciary. After approval by the National Assembly, the PDG Central Committee, and the President, the Assembly unanimously adopted the constitution in March 1991. Multiparty legislative elections were held in 1990-91, despite the fact that opposition parties had not been declared formally legal. In spite of this, the elections produced the first representative, multiparty National Assembly. In January 1991, the Assembly passed by unanimous vote a law governing the legalization of opposition parties.[5]

After President Omar Bongo was re-elected in a disputed presidential election in 1993 with 51% of votes cast, social and political disturbances led to the 1994 Paris Conference and Accords, which provided a framework for the next elections. Local and legislative elections were delayed until 1996-97. In 1997, constitutional amendments put forward years earlier were adopted to create the Senate and the position of vice president, as well as to extend the president's term to 7 years.[5]

In October 2009, newly-elected President Ali Bongo Ondimba began efforts to streamline the government. He eliminated 17 minister-level positions. He also abolished the vice president position and reorganized the portfolios of numerous ministries, bureaus, and directorates with the intention of reducing corruption and government bloat. In November 2009, President Bongo Ondimba announced a new vision for the modernization of Gabon, called "Gabon Emergent." This program contains three pillars: Green Gabon, Service Gabon, and Industrial Gabon. The goals of Gabon Emergent are to diversify the economy so that Gabon becomes less reliant on petroleum, to eliminate corruption, and to modernize the workforce. Under this program, exports of raw timber have been banned, a government-wide census was held, the work day has been changed to eliminate a long midday break, and a national oil company was created.[5]

For administrative purposes, Gabon is divided into 9 provinces, which are further divided into 36 prefectures and 8 separate subprefectures. The president appoints the provincial governors, the prefects, and the subprefects.[5]

In provisional results, the ruling Gabonese Democratic Party (PDG) won 84 out of 120 parliamentary seats.

Gabon has a small, professional military of about 5,000 personnel, divided into army, navy, air force, gendarmerie, and national police. Gabonese forces are oriented to the defense of the country and have not been trained for an offensive role. A 1,800-member guard provides security for the president.[5]

Since independence, Gabon has followed a nonaligned policy, advocating dialogue in international affairs and recognizing each side of divided countries. In inter-African affairs, Gabon espouses development by evolution rather than revolution and favors regulated free enterprise as the system most likely to promote rapid economic growth. Gabon played an important leadership role in the stability of Central Africa through involvement in mediation efforts in Chad, the Central African Republic, Angola, the Republic of the Congo, the Democratic Republic of the Congo (D.R.C.), and Burundi. In December 1999, through the mediation efforts of President Bongo, a peace accord was signed in the Republic of the Congo (Brazzaville) between the government and most leaders of an armed rebellion. President Bongo was also involved in the continuing D.R.C. peace process, and played a role in mediating the crisis in Côte d'Ivoire. Gabonese armed forces were also an integral part of the Central African Economic and Monetary Community (CEMAC) mission to the Central African Republic. Gabon is a member of the United Nations (UN) and some of its specialized and related agencies, as well as of the World Bank; the IMF; the African Union (AU); the Central African Customs Union/Central African Economic and Monetary Community (UDEAC/CEMAC); EU/ACP association under the Lome Convention; the Communaute Financiere Africaine (CFA); the Organization of the Islamic Conference (OIC); the Nonaligned Movement; and the Economic Community of Central African States (ECCAS/CEEAC), among others. In 1995, Gabon withdrew from the Organization of Petroleum Exporting Countries (OPEC). Gabon was elected to a non-permanent seat on the UN Security Council for January 2010 through December 2011 and held the rotating presidency in March 2010.[5]

On January 25, 2011, opposition leader André Mba Obame claimed the presidency, saying the country should be run by someone the people really wanted. He also selected 19 ministers for his government, and the entire group, along with hundreds of others, spent the night at United Nations headquarters. On January 26, the government dissolved Mba Obame's party. African Union chairman Jean Ping said that Mba Obame's action "hurts the integrity of legitimate institutions and also endangers the peace, the security and the stability of Gabon."[6] Interior Minister Jean Francois Ndongou accused Mba Obame and his supporters of treason.[6] The UN Secretary-General, Ban Ki-moon, said that he recognized Ondimba as the only official Gabonese president.

Provinces and departments

Gabon is divided into nine provinces and further divided into 37 departments.

The provinces are:

1. Estuaire
2. Haut-Ogooué
3. Moyen-Ogooué
4. Ngounié
5. Nyanga
6. Ogooué-Ivindo
7. Ogooué-Lolo
8. Ogooué-Maritime
9. Woleu-Ntem

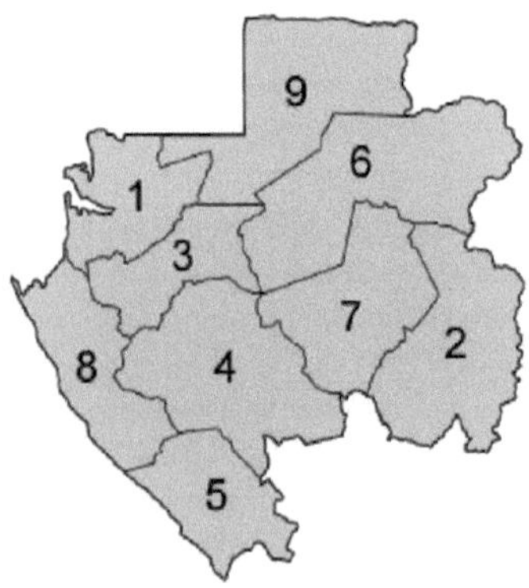

Provinces of Gabon

Geography

Gabon is located on the Atlantic coast of central Africa. Located on the equator, between latitudes 3°N and 4°S, and longitudes 8° and 15°E. Gabon generally has an equatorial climate with an extensive system of rainforests covering 85% of the country. There are three distinct regions: the coastal plains (ranging between 20 to 300 km from the ocean's shore), the mountains (the Cristal Mountains to the northeast of Libreville, the Chaillu Massif in the centre, culminating at 1575 m with Mont Iboundji), and the savanna in the east. The coastal plains form a large section of the World Wildlife Fund's Atlantic Equatorial coastal forests ecoregion and contain patches of Central African mangroves especially on the Muni River estuary on the border with Equatorial Guinea.

Satellite image of Gabon

Gabon's largest river is the Ogooué which is 1200 km long. Gabon has three karst areas where there are hundreds of caves located in the dolomite and limestone rocks. Some of the caves include Grotte du Lastoursville, Grotte du Lebamba, Grotte du Bongolo, and Grotte du Kessipougou. Many caves have not been explored yet. A National Geographic Expedition visited the caves in the summer of 2008 to document them (Expedition Website [7]).

Gabon is also noted for efforts to preserve the natural environment. In 2002, President Omar Bongo Ondimba put Gabon firmly on the map as an important future ecotourism destination by designating more than 11% of the nation's territory to be part of its national park system (with 13 parks in total), one of the largest proportions of nature parkland in the world. Natural resources include: petroleum, magnesium, iron, gold, uranium, and forests.

Economy

Gabon's economy is dominated by oil. Oil revenues comprise roughly 46% of the government's budget, 43% of gross domestic product (GDP), and 81% of exports. Oil production is now declining rapidly from its high point of 370,000 barrels per day in 1997. Some estimates suggest that Gabonese oil will be expended by 2025. In spite of the decreasing oil revenues, planning is only now beginning for an after-oil scenario.[5]

Gabonese public expenditures from the years of significant oil revenues were not spent efficiently. Overspending on the Transgabonais railroad, the oil price shock of 1986, the CFA franc devaluation of 1994, and low oil prices in the late 1990s caused serious debt problems that still plague the country.[5]

Gabon earned a poor reputation with the Paris Club and the International Monetary Fund (IMF) over the management of its debt and revenues. Successive IMF missions have criticized the government for overspending on off-budget items (in good years and bad), over-borrowing from the Central Bank, and slipping on the schedule for privatization and administrative reform. However, in September 2005, Gabon successfully concluded a 15-month Stand-By Arrangement with the IMF. Another 3-year Stand-By Arrangement with the IMF was approved in May 2007. Because of the financial crisis and social developments surrounding the death of President Omar Bongo and the elections, Gabon was unable to meet its economic goals under the Stand-By Arrangement in 2009. Negotiations with the IMF are ongoing.[5]

Gabon's oil revenues have given it a strong per capita GDP of $8,600, extremely high for the region. On the other hand, a skewed income distribution and poor social indicators misrepresent the situation if only GDP is taken into account. The richest 20% of the population receive over 90% of the income while about a third of all Gabonese live in poverty.[5]

The economy is highly dependent on extraction of abundant primary materials. Prior to the discovery of oil, logging was the pillar of the Gabonese economy. Today, logging and manganese mining are the other major income generators. Recent explorations point to the presence of the world's largest unexploited iron ore deposit. For many living in the countryside without access to employment in extractive industries, remittances from family members in urban areas or subsistence activities provide income.[5]

Many foreign and local observers have consistently lamented the lack of diversity in the Gabonese economy. Various factors have so far stymied additional industries—a small market of about 1 million people, dependence on French imports, inability to capitalize on regional markets, lack of entrepreneurial zeal among the Gabonese, and the fairly regular stream of oil "rent". Further investment in agricultural or tourism sectors is complicated by poor infrastructure. The small processing and service sectors that do exist are largely dominated by a few prominent local investors.[5]

At World Bank and IMF insistence, the government embarked in the 1990s on a program of privatization of its state-owned companies and administrative reform, including reducing public sector employment and salary growth, but progress has been slow. The new government has voiced a commitment to work toward an economic transformation of the country but faces significant challenges to realize this goal.[5]

Demographics

Gabon has a population that is estimated at 1,545,255.[5] Historical and environmental factors caused Gabon's population to decline between 1900 and 1940. It has one of the lowest population densities of any country in Africa,[5] and the highest Human Development Index in Sub-Saharan Africa.[4]

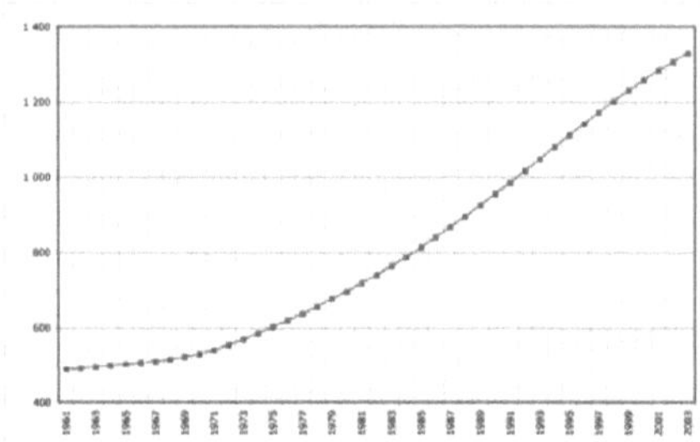

Population, in thousands, of Gabon from 1961 to 2003

Almost all Gabonese are of Bantu origin, though Gabon has at least forty ethnic groups with diverse languages and cultures.[5] The Fang are generally thought to be the largest,[5] although recent census data seem to favor the Nzebi. Others include the Myene, Kota, Shira, Puru, and Kande.[5] Ethnic boundaries are less sharply drawn in Gabon than elsewhere in Africa. There are also various Pygmy peoples: the Bongo, Kota, and Baka; the latter speak the only non-Bantu language in Gabon.

Most ethnicities are spread throughout Gabon, leading to constant contact and interaction among the groups. Intermarriage between the ethnicities is quite common, helping reduce ethnic tensions. French, the official language, is a unifying force. The Democratic Party of Gabon (PDG)'s historical dominance also has served to unite various ethnicities and local interests into a larger whole. More than 10,000 native French live in Gabon, including an estimated 2,000 dual nationals.[5] It is estimated that 80% of the country's population are able to speak French, and that 30% of Libreville residents are native speakers of the language. Nationally, 32% of the Gabonese people speak the Fang language as a mother tongue.[8]

Major religions practiced in Gabon include Christianity (Roman Catholicism and Protestantism), Bwiti, Islam, and indigenous animistic religion.[9] Many persons practice elements of both Christianity and traditional indigenous religious beliefs.[9] Approximately 73 percent of the population, including noncitizens, practice at least some elements of Christianity, including the syncretistic Bwiti; 12 percent practice Islam (of whom 80 to 90 percent are foreigners); 10 percent practice traditional indigenous religious beliefs exclusively; and 5 percent practice no religion or are atheists.[9] Gabon's literacy rate is 63.2%.[10]

Culture and media

Gabonese mask

Gabonese music is lesser-known in comparison with regional giants like the Democratic Republic of the Congo and Cameroon. The country boasts an array of folk styles, as well as pop stars like Patience Dabany and Annie Flore Batchiellilys, a Gabonese singer and renowned live performer. Also known are guitarists like Georges Oyendze, La Rose Mbadou and Sylvain Avara, and the singer Oliver N'Goma. Imported rock and hip hop from the US and UK are popular in Gabon, as are rumba, makossa and soukous. Gabonese folk instruments include the obala, the ngombi, balafon and traditional drums.

A country with a primarily oral tradition up until the spread of literacy in the 21st century, Gabon is rich in folklore and mythology. "Raconteurs" are currently working to keep traditions alive such as the mvett among the Fangs and the ingwala among the Nzebis.

Gabon also features internationally celebrated masks, such as the n'goltang (Fang) and the relicary figures of the Kota. Each group has its own set of masks used for various reasons. They are mostly used in traditional ceremonies such as marriage, birth and funerals. Traditionalists mainly work with rare local woods and other precious materials.

Radio-Diffusion Télévision Gabonaise (RTG), which is owned and operated by the government, broadcasts in French and indigenous languages. Color television broadcasts have been introduced in major cities. In 1981, a commercial radio station, Africa No. 1, began operations. The most powerful radio station on the continent, it has participation from the French and Gabonese governments and private European media. In 2004, the government operated two radio stations and another seven were privately owned. There were also two government television stations and four privately owned. In 2003, there were an estimated 488 radios and 308 television sets for every 1,000 people. About 11.5 of every 1,000 people were cable subscribers. Also in 2003, there were 22.4 personal computers for every 1,000 people and 26 of every 1,000 people had access to the Internet. The national press service is the Gabonese Press Agency, which publishes a daily paper, Gabon-Matin (circulation 18,000 as of 2002). L'Union in Libreville, the government-controlled daily newspaper, had an average daily circulation of 40,000 in 2002. The weekly Gabon d'Aujourdhui, is published by the Ministry of Communications. There are about nine privately owned periodicals which are either independent or affiliated with political parties. These publish in small numbers and are often delayed by financial constraints. The constitution of Gabon provides for free speech and a free press, and the government supports these rights. Several periodicals actively criticize the government and foreign publications are widely available.

Health

Most of the health services of Gabon are public, but there are some private institutions, of which the best known is the hospital established in 1913 in Lambaréné by Albert Schweitzer. Gabon's medical infrastructure is considered one of the best in West Africa. By 1985 there were 28 hospitals, 87 medical centers, and 312 infirmaries and dispensaries. As of 2004, there were an estimated 29 physicians per 100,000 people. Approximately 90% of the population had access to health care services. In 2000, 70% of the population had access to safe drinking water and 21% had adequate sanitation. A comprehensive government health program treats such diseases as leprosy, sleeping sickness, malaria, filariasis, intestinal worms, and tuberculosis. Rates for immunization of children under the age of one were 97% for tuberculosis and 65% for polio. Immunization rates for DPT and measles were 37% and 56% respectively. Gabon has a domestic supply of pharmaceuticals from a large, modern factory in Libreville. The total fertility rate has decreased from 5.8 in 1960 to 4.2 children per mother during childbearing years in 2000. Ten percent of all births were low birth weight. The maternal mortality rate was 520 per 100,000 live births as of 1998. In

2005, the infant mortality rate was 55.35 per 1,000 live births and life expectancy was 55.02 years. As of 2002, the overall mortality rate was estimated at 17.6 per 1,000 inhabitants. The HIV/AIDS prevalence was 8.10 per 100 adults in 2003. As of 2004, there were approximately 48,000 people living with HIV/AIDS in the country. There were an estimated 3,000 deaths from AIDS in 2003.

Education

Gabon's education system is regulated by two Ministries: The Ministry of Education, which is in charge of pre-kindergarten through the last High School Grade, and the Ministry of Higher Education and Innovative Technologies, which is in charge of Universities and Higher Education and Professional Schools.

Education is compulsory for children ages 6 to 16 years under the Education Act. Most children in Gabon start their school lives by attending Nurseries or "Crèche", then Kindergarten known as "Jardins d'Enfants". At age 6, they are enrolled in Primary School, "École Primaire" which is made up of six grades. The next level is "École Secondaire", which is made up of seven grades. The planned graduation age is 19 years old. Those who graduate can apply for Universities or institutions of Higher learning, such as engineering schools or business schools.

The government has used oil revenue for school construction, paying teachers' salaries, and promoting education, including in rural areas. However, maintenance of school structures, as well as teachers' salaries, has been declining. In 2002 the gross primary enrollment rate was 132 percent, and in 2000 the net primary enrollment rate was 78 percent. Gross and net enrollment ratios are based on the number of students formally registered in primary school and therefore do not necessarily reflect actual school attendance. As of 2001, 69 percent of children who started primary school were likely to reach grade 5. Problems in the education system include poor management and planning, lack of oversight, poorly qualified teachers, and overcrowded classrooms.[11]

See also

- Transport in Gabon

References

[1] Department of Economic and Social Affairs Population Division (2009) (PDF). *World Population Prospects, Table A.1* (http://www.un.org/esa/population/publications/wpp2008/wpp2008_text_tables.pdf). 2008 revision. United Nations. . Retrieved 2009-03-12.

[2] "Gabon" (http://www.imf.org/external/pubs/ft/weo/2011/01/weodata/weorept.aspx?sy=2008&ey=2011&scsm=1&ssd=1&sort=country&ds=.&br=1&c=646&s=NGDPD,NGDPDPC,PPPGDP,PPPPC,LP&grp=0&a=&pr.x=28&pr.y=1). International Monetary Fund. . Retrieved 2011-04-21.

[3] "Human Development Report 2010" (http://hdr.undp.org/en/media/HDR_2010_EN_Table1.pdf). United Nations. 2010. . Retrieved 5 November 2010.

[4] "Human Development Indices: A statistical update 2010 - HDI rankings" (http://hdr.ndp.org/en/statistics/). *Human Development Reports*. United Nations Development Programme. 2010. . Retrieved 2011-08-17.

[5] Background note: Gabon (http://www.state.gov/r/pa/ei/bgn/2826.htm). U.S. Department of State (August 4, 2010). *This article incorporates text from this source, which is in the public domain.*

[6] Goma, Yves Laurent (2011-01-26). "Gabon opposition leader declares himself president" (http://hosted.ap.org/dynamic/stories/A/AF_GABON?SITE=NCWIN&SECTION=HOME&TEMPLATE=DEFAULT). *Winston-Salem Journal*. Associated Press. . Retrieved 2011-01-26.

[7] http://blogs.nationalgeographic.com/blogs/gabon-caves

[8] Conrad Ouellon. "Le Gabon" (http://www.tlfq.ulaval.ca/axl/afrique/gabon.htm). Tlfq.ulaval.ca. . Retrieved 2010-06-27.

[9] International Religious Freedom Report 2007: Gabon (http://www.state.gov/g/drl/rls/irf/2007/90098.htm). United States Bureau of Democracy, Human Rights and Labor (September 14, 2007). *This article incorporates text from this source, which is in the public domain.*

[10] CIA - The World Factbook - Gabon (https://www.cia.gov/library/publications/the-world-factbook/geos/gb.html)

[11] "Gabon". *2005 Findings on the Worst Forms of Child Labor* (http://www.dol.gov/ilab/media/reports/iclp/tda2005/tda2005.pdf). Bureau of International Labor Affairs, U.S. Department of Labor (2006). *This article incorporates text from this source, which is in the public domain.*

Further reading

- Ghazvinian, John (2008). *Untapped: The Scramble for Africa's Oil*. Orlando: Harcourt. ISBN 0151011389.
- Petringa, Maria (2006). *Brazza, A Life for Africa*. Bloomington, IN: AuthorHouse. ISBN 1425911986.
- Rich, Jeremy (2007). *A Workman Is Worthy of His Meat: Food and Colonialism in the Gabon Estuary*. Lincoln: University of Nebraska Press. ISBN 0803207417.
- Shaxson, Nicholas (2007). *Poisoned Wells: The Dirty Politics of African Oil*. New York: Palgrave Macmillan. ISBN 1403971943.
- Warne, Sophie (2003). *Bradt Travel Guide: Gabon and São Tomé and Príncipe*. Guilford, CT: Chalfont St. Peter. ISBN 1841620734.
- Yates, Douglas A. (1996). *The Rentier State in Africa: Oil Rent Dependency and Neo-colonialism in the Republic of Gabon*. Trenton, NJ: Africa World Press. ISBN 0865435200.

External links

- Official website (http://http://www.legabon.org/)
- Gabon travel guide from Wikitravel
- Virtual Tour of the National Parks (http://www.gabon-nature.com/)
- Gabon (https://www.cia.gov/library/publications/the-world-factbook/geos/gb.html) entry at *The World Factbook*
- Gabon (http://www.dmoz.org/Regional/Africa/Gabon/) at the Open Directory Project
- Wikimedia Atlas of Gabon
- Gabon travel guide from Wikitravel

bjn:Gabon gag:Gabon mrj:Габон

Article Sources and Contributors

Mining industry of Equatorial Guinea *Source*: http://en.wikipedia.org/w/index.php?title=Mining_industry_of_Equatorial_Guinea *Contributors*: Circeus, ErikD, Hugo999, Kelapstick, Manxruler, MisfitToys, Neutrality, OpenToppedBus, Ottre, Piledhigheranddeeper, Poptart87, Rjhatl, Rjwilmsi, 1 anonymous edits

Equatorial Guinea *Source*: http://en.wikipedia.org/w/index.php?title=Equatorial_Guinea *Contributors*: *drew, -- April, 10metreh, 172, 23skidoo, A Werewolf, A-giau, ABF, AVM, Aaker, Abhijitsathe, Abhishekbh, Acntx, Acs4b, AdjustShift, Adrianbrodal, Aeusoes1, Aezram, Aflis, Againme, Aitias, Akanemoto, Alan Liefting, AlefZet, Alex45, Alexsoddy, Allstarecho, Am86, Amakuru, Amire80, Anadine, Andy Marchbanks, Antiuser, AntoinOnline, Aotearoa, ArchonMeld, Arjayay, Arniep, Arthena, Asterion, Attilios, Aude, Audiomecanik, Australia boy, Av9, Avala, Avs5221, AwamerT, Ayene, Bab-a-lot, Backpack, Baronnet, BarretBonden, Bart133, Bayerischermann, Bazonka, Beiner16, Beliar, Belovedfreak, Bemoeial, Ben Ben, Benge, Bestpay, Bfan1824, Big Adamsky, BigBen21, Bigturnip, Bkell, BlackOcelot, Blessthishouse, Bluedenim, Bob1larry, Bobo192, Boing! said Zebedee, Brian0918, Buaidh, Bullzeye, CJBR, Califate123!, CambridgeBayWeather, Canadian-Bacon, Cantus, Caponer, Cardiox, CarlKenner, Cburnett, Cdc, Cecil, Ceeded, Checco24, Check two you, Checkmate987, ChongDae, Chris the speller, Château-mousse, CieloEstrellado, Ckafura, Clogs13, Cnyborg, Coinmanj, Consuleg, Conte di Cavour, Conversion script, Craigy144, Credema, Crimson Observer, Cristixav, Cwolfsheep, Cwsavage78, Cyberbrain666666, Cybercobra, D Monack, D6, DabMachine, Daduzi, Dan100, DanMS, Danger, Daniel Medina, Danny, Darwinek, David Kernow, Davidme, Deagle AP, Deanshan, Debresser, Delldot, Demicx, Denisarona, Deucalionite, Dharmabum420, Diannaa, Dimension31, Discospinster, Discott, Diwas, Djgrasso, Dlohcierekim, Doc Quintana, Docu, Dodiad, Domaleixo, Dominic, Dominus, Doulos Christos, Downwards, Drbug, Droll, Ds77, Dumboinround, E Pluribus Anthony, EGPR, ERcheck, Earl CG, Eastlaw, Edge222, Edward, El C, Electionworld, Eleman, Ellmist, Elostirion, Enlil Ninlil, Enriquemartino, Entheta, Epbr123, Ermacrules29, Error, EugeneZelenko, Everyking, Evlekis, Extra999, Ezeu, FCYTravis, FMPARENTE, Facteditor, Fdt, Feydey, Fiftyquid, Filiep, FilipeS, Firth m, Fishal, Fivemack, Flatterworld, Foofbun, Francisco Valverde, Freewayuk, Funnyhat, Fuper, Fæ, Gabbe, GabrieL, Gaius Cornelius, GameLegend, Gauss, Gbleem, Geni, Gggh, Gilliam, Gimferrer, Goalie16, Godhowdoifindaname?, Golbez, Good Olfactory, Gr8opinionater, Green Giant, Grendelkhan, Ground Zero, Gryffindor, Guillaume-Sabourin, Gzkn, Gznorneplatz, Gzornenplatz, HPBarreto, Habitmeans383883, Haham hanuka, HappyInGeneral, Haukurth, HeartofaDog, Heegoop, HeikoEvermann, Heitordp, Henry Dorsett Case, Herostratus, HexaChord, Hiddekel, HisSpaceResearch, Hitssquad, Hjrtce30, Hottentot, Howcheng, Husond, Høst, Ian Pitchford, Ibbn, Ief, Immunize, Iol, Ipsumesse, Iridescent, IrishPete, IronGargoyle, Istanbuljohnm, It Is Me Here, It is me i think, J.delanoy, J04n, JHP, Jacob.jose, Jahangard, JamesBWatson, JamesR, JamesTseng, Jameslwoodward, Jamie C, Jazz457, Jcmenal, Jeanpmason, Jeff G., JeffreyN, Jeroen, Jiang, Jimmy Slade, Jimtaip, JinJian, Joanacaldas, Joao, Joe Schmedley, Joeldl, JohnArmagh, Jojit fb, JonC0001, JonHarder, JonSangster, Jorge Stolfi, Jorunn, Jose77, Joseph Solis in Australia, Josephabradshaw, Jotamar, Jullag, Jummai, Jusdafax, K4zem, KFP, Kahuzi, Kanags, Katpryde, Kelisi, Kenneth Shabby, Kevyn, Kifaya, Kintetsubuffalo, KirrVlad, Kman543210, Kman665, Knutux, Kolins, Kotniski, Koyaanis Qatsi, KoyaanisQatsi, Kralizec!, Kschwerdt514, Kvasir, Kwamikagami, Kwiki, Landrjm, Lesouris, LiDaobing, Liftarn, Lightmouse, LilHelpa, Lilac Soul, Limongi, Lincmad, Lucidx, Luis wiki, Luizdl, Luka Jačov, Luke the man o, MJCdetroit, MJD86, Madhero88, Magunz, Maksim L., Malhonen, Mancunius, Marek69, Markvs, Markyah, Maroux, Marzolian, Materialscientist, McDogm, McTrixie, Merijn2, Mesgul82, Meznaric, Mic, Middayexpress, Midemer, Minimac, MinnesotanConfederacy, Miquonranger03, Miranche, Mistakefinder, Mlc409, Modify, Morwen, MrYEAY55, Muhammad Daffa Rambe, Mustrum, MuzikJunky, Mwmonk, Naryathegreat, Natl1, NawlinWiki, Netoholic, Neutrality, Ngb, NicholasRWard, Nicksss93, Nightstallion, Nimzzs, Noctibus, Notheruser, Novem Linguae, NuclearWarfare, Numbo3, Nunquam Dormio, Nyttend, Oatmeal batman, Ohnoitsjamie, Oop, Osman22, OwenBlacker, PMDrive1061, Palica, Paploo, Paste, Pavel Vozenilek, Peter Horn, PeterisP, Petrux, Peyre, Pgan002, Picapica, Pigman, Pilim, PimRijkee, Pinethicket, Plasticup, Pmc4ever, Pointillist, Polaron, Polo aus85, PrestonH, Pridkett, Pristino, Proud Ho, Pteron, Puertorico1, Pwt898, Quest for Truth, Quuxplusone, R'n'B, Ramirez72, RandomP, Rarelibra, Ratfinx, Raymond Cruise, Rebturner, Recurring dreams, Red Winged Duck, Redfearnb, Redhill54, Reedy, Remus John Lupin, Res2216firestar, Rheinländer, Rich Farmbrough, Rick Block, Rickard Vogelberg, Rif Winfield, Rjhatl, Rjwilmsi, Rmosler2100, Rob5691, Robartinc, Robbiemuffin, RockMFR, Roke, Romanm, Romihaitza, Ronhjones, RookZERO, S. Neuman, SD6-Agent, SFC9394, Saforrest, Saleemhali, SamEV, Samantha555, Sarcelles, Savant13, Schaengel89, Scikidus, Scipius, Sciurinæ, Scohoust, Scorpion0422 Sucks, Scythian99, Seb az86556, Secfan, Seraphcrono, Sesel, Severo, Sfahey, Sgsilver, Shadowjams, Shanes, Shushruth, Sillyfolkboy, SimonP, Sirkeg, SisuUnlimited, Sivlve32, Sjolie2000, Sjwk, Skapur, Skinnyweed, Skrofler, Sligocki, Sluzzelin, Smack, Smkolins, Snocrates, Snowdog, SoWhy, Spudtater, SqueakBox, Stan Shebs, Starblind, Storm Rider, Supaman89, Surachit, Sven70, TFCforever, TUF-KAT, TXiKi, Tabletop, Takagithegreat, Tangerines, Tbhotch, Template namespace initialisation script, Tempodivalse, Tesi1700, Thaclient, The Ogre, The Person Who Is Strange, The Thing That Should Not Be, The Transhumanist, TheKMan, Thecurran, Thehelpfulone, Thomasmallen, Tibullus, Tiddly Tom, Tobby72, Tobias Conradi, Tom, Tom Radulovich, Tom harrison, Tom-b, Tonym88, TotallyTempo, Tpbradbury, Triberocker, Tterrag, Tukes, Ud terrorist, Uncle Dick, Urhixidur, Ursutraide, V6g3h7, Varlagas, Verdlanco, Vipgam, VityUvieu, VolatileChemical, VoodooIsland, Vrenator, Vssun, WACourson, WHATaintNOcountryIeverHEARDofDOtheySPEAKenglishINwhat, Waltloc, Wayne Slam, WhisperToMe, Wik, Wiki Wikardo, Wikiklrsc, Will Beback Auto, Willem Tijssen, Wimt, Wisdom89, Wizzy, Woodstein52, Woohookitty, X!, Xana1, Yabbadab, Yahia.barie, Ynhockey, Yug, Yvesnimmo, Zanimum, Zhou Yu, Ziggurat, Zscout370, Zzorse, Саша Стефановић, 973 anonymous edits

Economy of Equatorial Guinea *Source*: http://en.wikipedia.org/w/index.php?title=Economy_of_Equatorial_Guinea *Contributors*: Acntx, Alterboy, Andre Engels, Australia boy, Beagel, Bobblewik, Brammen, Briaboru, Dbachmann, Dermatas, Divoc, Donarreiskoffer, Edward, Esprit15d, Flatterworld, Fredbauder, Gene Nygaard, Gerbrant, Goodvac, Greenman, Japanese Searobin, Joao Xavier, Joseph Solis in Australia, Kashmiri, Kbdank71, Koavf, Kontar, Koyaanis Qatsi, Kralizec!, Kukini, Lightmouse, Lime82, LittleDan, Mamduh, Matt Deres, McTrixie, Natalya, Nishant ram2007, Olivier, Petri Krohn, Qero, Rich Farmbrough, Rich257, Robyvecchio, Sango123, SimonP, Sonett72, The Epopt, Thricecube, Topbanana, Ud terrorist, Urhixidur, Wapcaplet, Win.monroe, Wizzy, 23 anonymous edits

Gross national income *Source*: http://en.wikipedia.org/w/index.php?title=Gross_national_income *Contributors*: Actarux, Aequo, Alessin, Alex1011, Antonen, Arcenciel, AvicAWB, Bsimmons666, Bucketsofg, Cholmes75, Cristianrock2, Cybercobra, DeadEyeArrow, DirkvdM, Dreispt, Dúnadan, Epbr123, Espoo, Fryed-peach, GRBerry, Iridescent, J.delanoy, JamieA, John, John Quiggin, Joseph Solis in Australia, Junling, Klehti, Kozuch, Kwamikagami, Lochaber, MaCRoEco, Malleus Fatuorum, Modster, Narvalo, Oden, Onceonthisisland, Paul Mundt, Pearle, Ribbit, Rumping, Siwulek, Skillet5, TKD, Tangerines, Thingg, Tjl1128, Tomeasy, Tommy2010, Wavelength, Webergerald, Yakiv Gluck, 67 anonymous edits

Hydrocarbon exploration *Source*: http://en.wikipedia.org/w/index.php?title=Hydrocarbon_exploration *Contributors*: Ahmad510, Airondias, Ala.foum, Alexhard, Allen Info, Allen Mesch, Altafqadir, Arthena, Asher196, Austinbirdman, BJ Axel, Beagel, Beetstra, Berland, Billinghurst, Blaxthos, Bluemoose, Bob W Allen, Brettpaul2000, Btball, Casey56, Cgingold, DominikWells, Edward, Electromagnetic, Fluent aphasia, Frbloke101, FredR, Gaius Cornelius, Geologyguy, Gire 3pich2005, Glenn, IcycleMort, Jcpalma, Jncraton, Joseph Solis in Australia, Kelly Martin, LeadSongDog, LigerThai, Lisa-tms, Mbeychok, Mboverload, Mikenorton, Mwtoews, Noctibus, Oanjao, OrbitalPete, PETRSCIENT, PMeulbroek, Palosirkka, Peschle, Peterbmoorman, Pinethicket, Qfl247, RG2, Reach Out to the Truth, Ronchristie, Schmiteye, Seoss, Shiftchange, Sid zoom, SimonP, Sjakkalle, TastyCakes, Tchan012, Templetongore, Topbanana, TurtleShroom, Ufwuct, Veriss1, Versageek, Vsmith, Wakebrdkid, Wavelength, Webschem, WriterHound, Zero Gravitas, 75 anonymous edits

Spanish Empire *Source*: http://en.wikipedia.org/w/index.php?title=Spanish_Empire *Contributors*: 041744, 08OceanBeach SD, 1nyclove, 2, 23prootie, 5 albert square, AE.RUNAWAY.FE, AME67, Aaron Gibbons, Adam sk, Afinati, AgeofEmpire, Ahoerstemeier, Akerbeltz, Al-Iskandar Tzaraath, Alai, Alansohn, Albizia, Albrecht, Aldis90, Alensha, Alex S, Alexf, AlexiusHoratius, Alfonso Márquez, Alfonso", Alphachimp, Alrededor, Amillar, Andersmusician, Andrew Dalby, Angel ivanov angelov, Anonymous from the 21st century, Antipoeten, Apoivre, Art LaPella, Arthur Wellesley, Artorius, Asidemes, Asocall, Asterion, Astrotrain, Atoric, Attilios, Augicoins, AxelBoldt, BACbKA, Bab-a-lot, Bachrach44, Barticus88, Batistontain, Bcnviajero, Bcorr, Beardo, Beli.Lugh, Belligero, Bentaguayre, Berthaslidesdale, Beyond My Ken, Bill Thayer, BjornVDM, Bkell, Blanchardb, Blaxthos, Bluemask, Bobblehead, Bobbyb373, Bobo192, Bombadil1986, Boycottthecaf, BurgererSF, Byelf2007, C.Kent87, C.lettingaAV, CJLL Wright, CSvBibra, CWii, Califate123!, CapitalR, Chameleon, Chhoro, Chmod007, Choess, Chris Capoccia, Christophenstein, Closedmouth, Coffee, Colonies Chris, Commander Keane, CommonsDelinker, Computerjoe, Corriebertus, Cosialscastells, Ctalmer, Cuak's, Curtbob7, Cush, Cuvtixo, Cyfal, Cyrusc, D, DabMachine, DanielNuyu, Darklilac, Darkwind, Darwinek, Dbachmann, De Fernández, DeadEyeArrow, December12345678, Declangraham, Deibid, Delta 51, Dentren, DerHexer, Dj nix, Doctor Boogaloo, Docu, Domino theory, DoubleBlue, Dougweller, Draeco, Drmies, Drumguy8800, Duncan1800, Dunkedun, Duroy, Eathomas, Eboracum, Ecemaml, EdJohnston, Edconn, Eddiewalsh25, Edison, El C, El Mayimbe, Emiliojcp, Emyndil, Enok, Enric Naval, Enriquegoni, EoGuy, Epbr123, Ereunetes, Error, Escarlati, Eugene van der Pijll, EuroHistoryTeacher, Eyadhamid, EyeSerene, FayssalF, Federicoospina, Fkehar, Flaminglemons, Flockmeal, Fram, Fredrik, Fribbler, Frran, GabrielF, Gaius iulius caesar, Gdr, Gennarous, Gil Gamesh, Gimferrer, Git2010, Golbez, Goudzovski, Goustien, Graham87, Grin20, Ground Zero, Gryffindor, Guidales, Gurch, HCPUNXKID, Hajor, Haukurth, Hayden120, Hchc2009, Hiberniantears, Hmains, Hrvyfrd, IAMTrust, IANVS, IIIV, IP Address, Ian Pitchford, Icey, Ignaciogavira, Immunize, Infrogmation, Iridescent, IstvanWolf, It Is Me Here, J00tel, J04n, JCRB, JaGa, Jacek Kendysz, JamesAM, JamesReyes, Jayron32, Jbmurray, Jeff G., Jmabel, John K, Johnn Dorian, Jooler, Jorge Stolfi, Joseph Solis in Australia, Kallahan, Keilana, Kingpin13, Kirill Lokshin, Kloth, Kman543210, Koavf, Komischn, Kozuch, Kudz75, Kurrop, Kuru, Kurykh, Kusma, Kwamikagami, Kyng, LSLM, Laocoont, Lee D, Lemonade100, Leujohn, Lightmouse, Ligulem, Lilac Soul, Look2See1, Lorates, Lord Cornwallis, LovesMacs, Loyalist Cannons, MEGV, MER-C, Macintosh User, Magus732, Malema125, MarcosKikos, Marek69, Markvigil1, Marlowe, Matt.leeck, MattieTK, Mattis, Mauri96, MegaSloth, Mel Etitis, Mic, Michael Hardy, Miguel, Miguelzinho, Mike riversdale, Mikehelms, Mindspillage, Miq, Mmortal03, Mmxx, Mnemeson, Mowsbury, Mr pand, Ms2ger, N419BH, NYArtsnWords, Nakon, Nascar1996, Naylor182, Neurodivergent, Nicmeynic, NightFalcon90909, Nightmare X, Nihiltres, Nonexistant User, NorCalHistory, Nosering, Notinasnaid, Ntgalloway, Nuninho Martins, Nuno Tavares, Nv1962, Onofre Bouvila, Optimist on the run, OttomanReference, OwenBlacker, Paladinltd, Para, Parkjunwung, Paste, Pearle, PedroPVZ, Penelope37, Perceval, Persian Poet Gal, Peter Karlsen, Petronas, Pfly, Philip Trueman, Pigman, Pilotguy, Platewq, PoccilScript, Poetaris, Pol098, Prashanthns, Projanen, Proud dyke, Provocateur, Psmith, R0uge, Ramirez72, Rande M Sefowt, Random user 39849958, RandomCritic, RandomP, RavenRasputin, Razorflame, Rcsprinter123, Red Heathen, Red4tribe, RedWolf, Reinyday, RekishiEJ, Resvoluci, RexNL, Rich Farmbrough, Rickyrab, Robocooop, Roert42, Roketjack, Rubenmajor, Rydia, S@bre, SE7, SEWilco, SKopp, SMP, Sabresman24, Sam Hocevar, SamEV, SamuelTheGhost, Sandik, SandyGeorgia, Sapient Homo, Scapone, Schwyz, Sean D Martin, Shanel, Shauni, Shimmin, Shoaler, Shtove, Sijo Ripa, Sir Edgar, Skidbladnir, Snowolf, Soccergirl2210, Soccergoalie11, Sonyray, Spangineer, Sparoynav, Srnec, Steven J. Anderson, Storeye, Str1977, Sunquanliangxiuhao, Supersexyspacemonkey, SynergyBlades, T L Miles, TShilo12, Tabletop, Taiwanman1, Tbhotch, Techman224, Tercio28, TexasAndroid, The Ogre, The Red Hat of Pat Ferrick, Theda, Thenewestdoctorwho, Thomas Paine1776, Tobby72, Tpbradbury, Trainunion, Trasamundo, Travelbird, TriniMuñoz, Tristanb, UBeR, Uannis, Ulric1313, Uncle Dick, Unyoyega, User2004, VWPY, Valentinian, Venerock, Veracruzian, Versageek, Victor12, Vis-a-visconti, Vreundl8, Waggers, Walter Humala, Waltpohl, Wanstiller, Wavelength, WereSpielChequers, Wetman, Wik, Wiki alf, Wikitza, Woohookitty, WorldPeace1492, Wpedzich, Wtmitchell, XGustaX, XPTO, Xaverius, Xinese-v, Yamaguchi, YellowAssessmentMonkey, Yngvadottir, Yoss Montmorency, Yug, ZX81, ZeD unknown,

Zepheus, Zerostar 17, Zhukov1945, Zoicon5, Zzuuzz, 1111 anonymous edits

Mongomo *Source*: http://en.wikipedia.org/w/index.php?title=Mongomo *Contributors*: Acntx, Arcarius, Caerwine, Darwinek, Dr. Blofeld, Good Olfactory, Grutness, JavierMC, Jor, Mr Accountable, Namiba, Pietaster, Secretlondon, Stan Shebs, 4 anonymous edits

United States Geological Survey *Source*: http://en.wikipedia.org/w/index.php?title=United_States_Geological_Survey *Contributors*: 2D, 5ju989nfhs50, Afromayun, AjaxSmack, Almit39, Analogue Kid, Articnomad, Authalic, Awickert, Bkell, BlueCanoe, BraneJ, Californian Treehugger, Ceyockey, Che090572, Chris857, Chrisfow, Citynoise, Cleared as filed, Clindberg, Craigy144, Curtvprice, CyanFOX, Dahv2000, Dananderson, Davewild, DeWaine, Dgovoni, Doopokko, Dralwik, Duncan.france, Eastlaw, Eeekster, Eequor, Egil, Emijrp, Epolk, Ericniehoff, Evolt, Florian Adler, Flyguy649, Fratrep, Fusstritt für S'Wolf, G716, Gdjha, George100, Giftlite, Giraffedata, Gobonobo, GraemeL, Hard Raspy Sci, Howdybob, IkonicDeath, Interiot, Ismail ngr, Isomorphic, Ixfd64, J.delanoy, Jaknouse, JerryFriedman, Jim.henderson, Johnpacklambert, Jonathan Callahan, Joshuahicks, Jurema Oliveira, Kbdank71, Kevmin, Kulikovsky, Lars Washington, Lipothymia, Locutus, Lokifer, Malepheasant, Mapsax, Markbnine, Mav, Mikaey, MikeHydro, Mikenorton, Minesweeper, Mmxx, Mulad, Mwng, Mwtoews, Mxn, Neutrality, Ohnoitsjamie, Olegwiki, Onco p53, PaulHanson, Pearle, Plazak, Postdlf, Ppntori, Qfl247, Quadell, QuakeSim, Rara evans, Ratberta, Redjar, RexNL, Rich Farmbrough, Rick Block, Rjwilmsi, Rlhuffine, Rmhirsch49, Robert Horning, Rougher07, SPUI, Sardanaphalus, ScottHorvath, Sm8900, Sorsanmetsastaja, Spydrlink, SusanLesch, Tbonnie, Tedernst, Tedickey, Teratornis, The Red, TheGrappler, TheKMan, Thincat, Tholme, Tillman, Tolivero, Vanka5, Vidale, Vsmith, Vter4life, WAS 4.250, Wars, WhisperToMe, Whywhenwhohow, WinstonSmith, Woohookitty, Xnatedawgx, Ybbor, Zero Gravitas, Zyxw, रोहित रावत, 123 anonymous edits

Artisan *Source*: http://en.wikipedia.org/w/index.php?title=Artisan *Contributors*: 100110100, Agnessmith, Aitias, Ajd, Alansohn, Aldedogn, Amcaja, Amire80, Anaxial, Andycjp, ArtisanTony, Atlan, Balthazarduju, Bearcat, BillShurts, Briangotts, Btilm, CALR, Call me Bubba, Caminante62, Capricorn42, Carabinieri, CatherineMunro, CheeseForum, ChildofMidnight, Chris Murphy, Chris9086, Chuckiesdad, Clubmarx, Corvus cornix, Cote d'Azur, Dadude3320, Daniel is broken, David Sneek, Dgies, EHFOH, Ebyabe, Eekerz, Enviroboy, Erianna, Espoo, FMAFan1990, Faradayplank, Fayimora, Flewis, Fplay, Galloping Moses, Gary King, Gcm, Gem-fanat, Glenn, Golbez, Goodnightmush, H3tfi3ld, Helvetica, HexaChord, Indarjeet, Irpen, J.delanoy, JHunterJ, Jackhynes, Jackollie, Jakro64, Jameboy, Johan1298, Jpbowen, KVDP, Kamocat, Katalaveno, Kbh3rd, KnowledgeOfSelf, Kyle Maxwell, Lainagier, LeCire, LindsayH, Lisamh, Mandarax, Martarius, Martin2Reid, Martpol, Mclay1, Mejor Los Indios, Michael Hardy, Minna Sora no Shita, Modernist, MrChupon, Myanw, NeonMerlin, Noisy, Onceler, Pax:Vobiscum, Phileas, Philip Trueman, Phuzion, Piotrus, Pizza1512, Pollinator, Qatter, Radagast83, Ranveig, Revth, ScottW, Secretlondon, Sfnhltb, Shinooshike, Silverfish, Smith7777, Sparkit, Stomme, Suwatest, Sverre, Syuz, TantalumTelluride, Techman224, Tempodivalse, Teomusa, TheObtuseAngleOfDoom, Trusilver, Tz'unun, Un chien andalou, Userafw, Vmenkov, Waitak, Was a bee, Wayland, Welsh, WikiUserPedia, Zundark, Zzuuzz, 193 anonymous edits

Gabon *Source*: http://en.wikipedia.org/w/index.php?title=Gabon *Contributors*: -- April, 1297, A Werewolf, A-giau, Aaker, Abhijitsathe, Acntx, Adrianbrodal, Aeusoes1, Africa's Eden, Ahoerstemeier, Akadruid, Akanemoto, Alam82, Alan Liefting, Alansohn, AlefZet, Alphaman11, Amakuru, Andy Marchbanks, Aridd, Art LaPella, AtomicDragon, Aude, Autopilot, Ayt Buyafar, BCKILLa, Babbage, Bamelzer, Bansp, Barneyboo, Bazonka, Bemoeial, Bencherlite, Bender235, Berean Hunter, Bestofmed, Big Adamsky, Biruitorul, Blechnic, Bobblewik, Bobo192, BozMo, Brammen, Briaboru, Brian0918, Btball, Buaidh, Buttersandpip, CJLL Wright, CLW, Cactus.man, Calliopejen1, Can't sleep, clown will eat me, Cantus, Caponer, CardinalDan, Carewolf, CarlKenner, Catgut, Cburnett, Cdc, Chanheigeorge, Chipmunkdavis, Civvi, Cjbrownfield, CommonsDelinker, Conte di Cavour, Conversion script, Corban.makaya, Corpx, Corvus13, Credema, Crohnie, CryptoDerk, Ctbolt, Cub68134, Cuneytewrares, Cureden, Cwolfsheep, Cyde, D6, DD Ting, DVdm, DabMachine, DanArmiger, Danny, Darwinek, David Edgar, David Kernow, Deflective, Demicx, Deor, Dirkbb, Discospinster, DivineIntervention, Dixke, Docu, Domaleixo, Dooboa, Dr. Blofeld, Drbug, Dubhe.sk, Dude22685, E Pluribus Anthony, Eb1941rp, Editor2423, Editorofthewiki, Eduxedu, Ekokoi, El C, Electionworld, Eliz81, ElockidAlternate, Emonobus, Epbr123, Eu.stefan, Everyking, Ezeu, Farosdaughter, Finlay, Fixer88, Flapdragon, Flatterworld, Flyax, Footwarrior, Freshacconci, Fritzpoll, Fvincent, G. Capo, Gabbe, Gaboneco, Gadfium, Gallen is a gun, GameLegend, Gauss, GenuineMongol, Gggh, Gilliam, Girmitya, Godheval, Gogo Dodo, Good Olfactory, GoodOlRickyTicky2, Gotipe, Graham87, Green Giant, Greenman, Grendelkhan, Gryffindor, Guaka, Gurgleboy, Gznorneplatz, Gzornenplatz, Habitmeans383883, Harel Newman, HeikoEvermann, Herostratus, Highfields, Hjyt, HkCaGu, IJA, IRP, IW.HG, Idaltu, Ief, Instinct, Irishguy, IronGargoyle, Istanbuljohnm, J.delanoy, Jacob.jose, Jahangard, Jake Wartenberg, JamesR, Jamesdreherjr, Jaxl, Jaysweet, Jeff G., Jhendin, Jimtaip, JoSePh, Jojit fb, Jonathan.s.kt, Jorunn, Jose77, Joseph Solis in Australia, Joshua Pace, Jpatokal, Juliancolton, Kahuzi, Kaisershatner, Kanags, Kbh3rd, KevinCuddeback, Kingpin13, Kintetsubuffalo, Kman543210, Kmorozov, KnowledgeOfSelf, Knutux, Kotniski, Koyaanis Qatsi, Kralizec!, Kubigula, Kusma, Kwamikagami, Largoplazo, LcawteHuggle, LeilaniLad, Lenar, LiDaobing, Liftarn, Lightmouse, LilHelpa, Lilac Soul, Limongi, Ling.Nut, Llywrch, LonelyMarble, Loren.wilton, Lozeldafan, Lucasalvarez, Luis wiki, Luisa and Guido, Luvaca10, MC10, MECU, MJCdetroit, MSClaudiu, Magioladitis, Mahongue, Maksim L., Makua1010101010, Malcolmxl5, Mario Žamić, Maroux, Martin451, MartinezMD, Martinwilke1980, Martpol, Marzoregulus@hotmail.com, Maurreen, McDogm, Mein Schwanz ist hart, Mentifisto, Mephistophelian, Mesgul82, MesserWoland, Mic, Middayexpress, Mightymights, Minesweeper, Miranda545, Mistakefinder, Mitch1981, Montrealais, Mrbert, Mrdie, Mrgouaf, Muhammad Daffa Rambe, Mygerardromance, N2e, NGerda, NawlinWiki, NeilN, NellieBly, New4325, Nightstallion, Northumbrian, Not the maximum, Notheruser, Olivier, Orangemike, Orgel, Oxymoron83, PDH, Paddles, Pascal, Patrick-br, Pejman47, Pengo, Peter Horn, PeterisP, Petrux, Philosopher Torin, Picaballo, PimRijkee, Pinethicket, Pmyteh, Polaron, Populer208, Profoss, Pruneau, Pteron, Quantpole, Qutezuce, Qxz, RFerreira, Randyc, Rarelibra, Rctycoplay, Rebecca, Remus John Lupin, RexNL, Rich Farmbrough, Richard75, Rick Block, Rickard Vogelberg, Rjwilmsi, RobNS, Rochelle14, Roke, Romanm, SFC9394, Saalstin, Sam Blacketer, Samantha555, Sarahmon, Satori Son, Schaengel89, Scipius, Scythian1, Scythian99, Sdrawkcab, Seb az86556, Seba5618, Secfan, Shanes, Shoeofdeath, Shrekkie45, Shrike, Silverhorse, SimonA1986, SimonP, SirYoda, SisuUnlimited, SkeeteRX8, Skew-t, Sluzzelin, Smith2006, Snowdog, Soap, Some jerk on the Internet, Spencer, Stan Shebs, Steenjager, Steinbach, Stephenb, Stunetii, Supadawg, Svick, TUF-KAT, Taaruvot, Tabletop, Targeman, Tcncv, Template namespace initialisation script, The Thing That Should Not Be, The Transhumanist, Therequiembellishere, Thingg, Thricecube, Thrill0401, Tide rolls, Tobz1000, Tpbradbury, Tree Biting Conspiracy, Tristanb, Uncle Dick, Valentinian, Vardion, Vchimpanzee, Victor12, Vlma111, VoodooIsland, Voyagerfan5761, Vrenator, Warofdreams, Wayward, WhisperToMe, Whpq, Wik, WikHead, Wiki Roxor, Wikiacc, Willem Tijssen, William Avery, Woodbags, Writtenright, Ygoloxelfer, Ynhockey, Yvesnimmo, ZX81, Zedmelon, Ziggypup123, Zscout370, , 614 anonymous edits

Image Sources, Licenses and Contributors

File:Flag of Equatorial Guinea.svg *Source*: http://en.wikipedia.org/w/index.php?title=File:Flag_of_Equatorial_Guinea.svg *License*: unknown *Contributors*: Anime Addict AA, Duschgeldrache2, Emc2, Fry1989, Homo lupus, Klemen Kocjancic, Maks Stirlitz, Mattes, Neq00, NeverDoING, Nightstallion, OAlexander, Permjak, ThomasPusch, 1 anonymous edits

File:Coat_of_arms_of_Equatorial_Guinea.svg *Source*: http://en.wikipedia.org/w/index.php?title=File:Coat_of_arms_of_Equatorial_Guinea.svg *License*: unknown *Contributors*: User:Szczepan1990

File:Location_Equatorial_Guinea_AU_Africa.svg *Source*: http://en.wikipedia.org/w/index.php?title=File:Location_Equatorial_Guinea_AU_Africa.svg *License*: unknown *Contributors*: User:Alvaro1984 18

File:Increase2.svg *Source*: http://en.wikipedia.org/w/index.php?title=File:Increase2.svg *License*: unknown *Contributors*: Sarang

File:Equatorialguineamap.png *Source*: http://en.wikipedia.org/w/index.php?title=File:Equatorialguineamap.png *License*: unknown *Contributors*: Electionworld, Telim tor, Tonym88

File:Equatorial Guinea provinces numbered.png *Source*: http://en.wikipedia.org/w/index.php?title=File:Equatorial_Guinea_provinces_numbered.png *License*: unknown *Contributors*: User:Golbez

File:bubi children.jpg *Source*: http://en.wikipedia.org/w/index.php?title=File:Bubi_children.jpg *License*: unknown *Contributors*: Original uploader was Tukes at en.wikipedia

Image:PD-icon.svg *Source*: http://en.wikipedia.org/w/index.php?title=File:PD-icon.svg *License*: unknown *Contributors*: Alex.muller, Anomie, Anonymous Dissident, CBM, MBisanz, Quadell, Rocket000, Strangerer, Timotheus Canens, 1 anonymous edits

File:Flag of the United States.svg *Source*: http://en.wikipedia.org/w/index.php?title=File:Flag_of_the_United_States.svg *License*: unknown *Contributors*: Anomie

File:Flag of the People's Republic of China.svg *Source*: http://en.wikipedia.org/w/index.php?title=File:Flag_of_the_People's_Republic_of_China.svg *License*: unknown *Contributors*: User:Denelson83, User:SKopp, User:Shizhao, User:Zscout370

File:Flag of Japan.svg *Source*: http://en.wikipedia.org/w/index.php?title=File:Flag_of_Japan.svg *License*: unknown *Contributors*: Anomie

File:Flag of Germany.svg *Source*: http://en.wikipedia.org/w/index.php?title=File:Flag_of_Germany.svg *License*: unknown *Contributors*: Anomie

File:Flag of France.svg *Source*: http://en.wikipedia.org/w/index.php?title=File:Flag_of_France.svg *License*: unknown *Contributors*: Anomie

File:Flag of the United Kingdom.svg *Source*: http://en.wikipedia.org/w/index.php?title=File:Flag_of_the_United_Kingdom.svg *License*: unknown *Contributors*: Anomie, Mifter

File:Flag of Italy.svg *Source*: http://en.wikipedia.org/w/index.php?title=File:Flag_of_Italy.svg *License*: unknown *Contributors*: Anomie

File:Flag of Brazil.svg *Source*: http://en.wikipedia.org/w/index.php?title=File:Flag_of_Brazil.svg *License*: unknown *Contributors*: Anomie

File:Flag of Spain.svg *Source*: http://en.wikipedia.org/w/index.php?title=File:Flag_of_Spain.svg *License*: unknown *Contributors*: Anomie

File:Flag of India.svg *Source*: http://en.wikipedia.org/w/index.php?title=File:Flag_of_India.svg *License*: unknown *Contributors*: Anomie, Mifter

File:Flag of Canada.svg *Source*: http://en.wikipedia.org/w/index.php?title=File:Flag_of_Canada.svg *License*: unknown *Contributors*: Anomie

Image:Erdöl Bohrturm.jpg *Source*: http://en.wikipedia.org/w/index.php?title=File:Erdöl_Bohrturm.jpg *License*: unknown *Contributors*: Brudersohn

Image:Mudlogging.JPG *Source*: http://en.wikipedia.org/w/index.php?title=File:Mudlogging.JPG *License*: unknown *Contributors*: User:Citypeek

File:Flag of New Spain.svg *Source*: http://en.wikipedia.org/w/index.php?title=File:Flag_of_New_Spain.svg *License*: unknown *Contributors*: User:Ningyou

File:Spanish Empire Anachronous 0.PNG *Source*: http://en.wikipedia.org/w/index.php?title=File:Spanish_Empire_Anachronous_0.PNG *License*: unknown *Contributors*: User:Trasamundo

File:Christopher Columbus9.jpg *Source*: http://en.wikipedia.org/w/index.php?title=File:Christopher_Columbus9.jpg *License*: unknown *Contributors*: ADGE, Balbo, Dodo, Prost666, Pullus In Fabula, Siebrand, Timichal, 1 anonymous edits

Image:The Death of Gaston de Foix in the Battle of Ravenna.jpg *Source*: http://en.wikipedia.org/w/index.php?title=File:The_Death_of_Gaston_de_Foix_in_the_Battle_of_Ravenna.jpg *License*: unknown *Contributors*: Diomede, G.dallorto, Kilom691, Kirill Lokshin, Shakko

Image:Columnas Plus Ultra.png *Source*: http://en.wikipedia.org/w/index.php?title=File:Columnas_Plus_Ultra.png *License*: unknown *Contributors*: User:Ignaciogavira

Image:Battle of Pavia.jpg *Source*: http://en.wikipedia.org/w/index.php?title=File:Battle_of_Pavia.jpg *License*: unknown *Contributors*: Alexandrin, Bibi Saint-Pol, DenghiùComm, FAR, G.dallorto, Henrytow, Kirill Lokshin, LimoWreck, Shizhao, Sidonius, Thib Phil, 2 anonymous edits

Image:Habsburg Map 1547.jpg *Source*: http://en.wikipedia.org/w/index.php?title=File:Habsburg_Map_1547.jpg *License*: unknown *Contributors*: edited by Sir Adolphus William Ward, G.W. Prothero, Sir Stanley Mordaunt Leathes

Image:Cateau-Cambresis.jpg *Source*: http://en.wikipedia.org/w/index.php?title=File:Cateau-Cambresis.jpg *License*: unknown *Contributors*: Cherubino, G.dallorto, MGA73, Pline, Razr, Shakko, Tangopaso, 4 anonymous edits

Image:Veen01.jpg *Source*: http://en.wikipedia.org/w/index.php?title=File:Veen01.jpg *License*: unknown *Contributors*: Otto van Veen

Image:Defensa de Cádiz contra los ingleses.jpg *Source*: http://en.wikipedia.org/w/index.php?title=File:Defensa_de_Cádiz_contra_los_ingleses.jpg *License*: unknown *Contributors*: Francisco de Zurbarán

Image:Invincible Armada.jpg *Source*: http://en.wikipedia.org/w/index.php?title=File:Invincible_Armada.jpg *License*: unknown *Contributors*: English School, 16th century

Image:Velazquez-The Surrenderof Breda.jpg *Source*: http://en.wikipedia.org/w/index.php?title=File:Velazquez-The_Surrenderof_Breda.jpg *License*: unknown *Contributors*: Carlomorino, Gons, Interpretix, Kweniston, Mbdortmund, Melesse, Pieter Kuiper, Player 03, Postdlf

Image:Battle of Nordlingen.jpg *Source*: http://en.wikipedia.org/w/index.php?title=File:Battle_of_Nordlingen.jpg *License*: unknown *Contributors*: AndreasPraefcke, DaBroMfld, GeorgHH, Richy, Wst, 9 anonymous edits

Image:Rocroi.jpg *Source*: http://en.wikipedia.org/w/index.php?title=File:Rocroi.jpg *License*: unknown *Contributors*: Krun, Olivier2, Pline, Rythin, Schmelzle, 2 anonymous edits

Image:Battle of Lepanto 1571.jpg *Source*: http://en.wikipedia.org/w/index.php?title=File:Battle_of_Lepanto_1571.jpg *License*: unknown *Contributors*: anonymous

Image:Inca-Spanish confrontation.JPG *Source*: http://en.wikipedia.org/w/index.php?title=File:Inca-Spanish_confrontation.JPG *License*: unknown *Contributors*: Lupo Original uploader was Dynamax at en.wikipedia

Image:Pedro De Valdivia Siedlungen in Chile 1540 bis 1553.jpg *Source*: http://en.wikipedia.org/w/index.php?title=File:Pedro_De_Valdivia_Siedlungen_in_Chile_1540_bis_1553.jpg *License*: unknown *Contributors*: User:WeHaKa

Image:Traite-Pyrenees.jpg *Source*: http://en.wikipedia.org/w/index.php?title=File:Traite-Pyrenees.jpg *License*: unknown *Contributors*: Laumosnier

Image:The Battle of Cape Passaro.jpg *Source*: http://en.wikipedia.org/w/index.php?title=File:The_Battle_of_Cape_Passaro.jpg *License*: unknown *Contributors*: User:Barbe-Noire

Image:FORTRESS OF SAN FELIPE CARTAGENA.JPG *Source*: http://en.wikipedia.org/w/index.php?title=File:FORTRESS_OF_SAN_FELIPE_CARTAGENA.JPG *License*: unknown *Contributors*: User:Federicoospina

Image:Spanish troops at Pensacola.jpg *Source*: http://en.wikipedia.org/w/index.php?title=File:Spanish_troops_at_Pensacola.jpg *License*: unknown *Contributors*: AnRo0002, Dodo, Editor at Large, GeorgHH, Infrogmation, KTo288, Mundo tarantino, Tony Rotondas, Urban, Valdavia, Verica Atrebatum, 8 anonymous edits

Image:Churruca death.jpg *Source*: http://en.wikipedia.org/w/index.php?title=File:Churruca_death.jpg *License*: unknown *Contributors*: Balbo, Erri4a, Ketamino, Kirill Lokshin, Makthorpe, Morio, Rama, Shakko, Shizhao, Xabier Armendaritz, 4 anonymous edits

Image:Francisco de Goya y Lucientes 026.jpg *Source*: http://en.wikipedia.org/w/index.php?title=File:Francisco_de_Goya_y_Lucientes_026.jpg *License*: unknown *Contributors*: AxelBoldt, Balbo, BomBom, Bukk, CommonsDelinker, Conscious, Diwas, Emijrp, Mattes, Paintman, Pitke, Sparkit, Thuresson, 4 anonymous edits

Image:Battle of Ayacucho.jpg *Source*: http://en.wikipedia.org/w/index.php?title=File:Battle_of_Ayacucho.jpg *License*: unknown *Contributors*: Adam Faanes, AnRo0002, Andres rojas22, Cookie, Olivier2, 1 anonymous edits

Image:Vizcayaexplodes.JPG *Source*: http://en.wikipedia.org/w/index.php?title=File:Vizcayaexplodes.JPG *License*: unknown *Contributors*: DO'Neil, Hugh Manatee, Pearle, Sfan00 IMG, Sreejithk2000, 1 anonymous edits

Image:Equatorialguineamap.png *Source*: http://en.wikipedia.org/w/index.php?title=File:Equatorialguineamap.png *License*: unknown *Contributors*: Electionworld, Telim tor, Tonym88

Image:ProtectoradoMarruecos.png *Source*: http://en.wikipedia.org/w/index.php?title=File:ProtectoradoMarruecos.png *License*: unknown *Contributors*: Ceresnet, Ecemaml, Ketamino, Kordas

Image:Catedral de lima2.jpg *Source*: http://en.wikipedia.org/w/index.php?title=File:Catedral_de_lima2.jpg *License*: unknown *Contributors*: Martin Garcia from Lima, Peru

file:Location map Equatorial Guinea main.png *Source*: http://en.wikipedia.org/w/index.php?title=File:Location_map_Equatorial_Guinea_main.png *License*: unknown *Contributors*: User:Dr. Blofeld

File:Red pog.svg *Source*: http://en.wikipedia.org/w/index.php?title=File:Red_pog.svg *License*: unknown *Contributors*: Anomie

file:US-GeologicalSurvey-Seal.svg *Source*: http://en.wikipedia.org/w/index.php?title=File:US-GeologicalSurvey-Seal.svg *License*: unknown *Contributors*: U.S. Government

file:USGS logo green.svg *Source*: http://en.wikipedia.org/w/index.php?title=File:USGS_logo_green.svg *License*: unknown *Contributors*: Mwtoews, Yarnalgo

Image:Clarence King.jpg *Source*: http://en.wikipedia.org/w/index.php?title=File:Clarence_King.jpg *License*: unknown *Contributors*: APPER, Clindberg, Denniss, MushiHoshiIshi, Zyxw

Image:Usgs-headquarters.jpeg *Source*: http://en.wikipedia.org/w/index.php?title=File:Usgs-headquarters.jpeg *License*: unknown *Contributors*: Clindberg, Joshuadavisphotography, WhisperToMe

Image:USGS Station.JPG *Source*: http://en.wikipedia.org/w/index.php?title=File:USGS_Station.JPG *License*: unknown *Contributors*: Original uploader was Analogue Kid at en.wikipedia

Image:Recent Earthquakes Last 8-30 Days.gif *Source*: http://en.wikipedia.org/w/index.php?title=File:Recent_Earthquakes_Last_8-30_Days.gif *License*: unknown *Contributors*: Julsep 6

Image:Bamun artisan.jpg *Source*: http://en.wikipedia.org/w/index.php?title=File:Bamun_artisan.jpg *License*: unknown *Contributors*: Amcaja, Bohème, RedWolf

File:Flag_of_Gabon.svg *Source*: http://en.wikipedia.org/w/index.php?title=File:Flag_of_Gabon.svg *License*: unknown *Contributors*: User:Gabbe, User:SKopp

File:Coat of arms of Gabon.svg *Source*: http://en.wikipedia.org/w/index.php?title=File:Coat_of_arms_of_Gabon.svg *License*: unknown *Contributors*: user:Herr Kriss, user:MesserWoland, user:Szczepan1990

File:Gabon (orthographic projection).svg *Source*: http://en.wikipedia.org/w/index.php?title=File:Gabon_(orthographic_projection).svg *License*: unknown *Contributors*: User:MSClaudiu

File:Loudspeaker.svg *Source*: http://en.wikipedia.org/w/index.php?title=File:Loudspeaker.svg *License*: unknown *Contributors*: Bayo, Gmaxwell, Husky, Iamunknown, Myself488, Nethac DIU, Omegatron, Rocket000, The Evil IP address, Wouterhagens, 11 anonymous edits

File:Flag of Gabon 1959-1960.svg *Source*: http://en.wikipedia.org/w/index.php?title=File:Flag_of_Gabon_1959-1960.svg *License*: unknown *Contributors*: User:Arz

File:Omar Bongo crop.jpg *Source*: http://en.wikipedia.org/w/index.php?title=File:Omar_Bongo_crop.jpg *License*: unknown *Contributors*: White house photo by Eric Draper

File:Gabon provinces numbered.png *Source*: http://en.wikipedia.org/w/index.php?title=File:Gabon_provinces_numbered.png *License*: unknown *Contributors*: User:Golbez

Image:Gabon_sat.png *Source*: http://en.wikipedia.org/w/index.php?title=File:Gabon_sat.png *License*: unknown *Contributors*: Cwolfsheep, Lokal Profil, Martin H.

File:Gabon demography.png *Source*: http://en.wikipedia.org/w/index.php?title=File:Gabon_demography.png *License*: unknown *Contributors*: Bender235, Demmo, Valérie75

File:TsogoMask.jpg *Source*: http://en.wikipedia.org/w/index.php?title=File:TsogoMask.jpg *License*: unknown *Contributors*: cliff1066

MIX
Papier aus verantwortungsvollen Quellen
Paper from responsible sources
FSC® C105338

Printed by Books on Demand GmbH, Norderstedt / Germany